Agricultural Communication

Editors

Professor V.K. Dubey

Ex. Head
Department of Extension Education
Institute of Agricultural Sciences
Banaras Hindu University
Varanasi (UP) India

Dr. J.P. Srivastava

Dean
Faculty of Agriculture
Chhatrapati Sahu Ji Maharaj University
Kanpur (UP) India

International Book Distributing Co.
(Publishing Division)

Published by
INTERNATIONAL BOOK DISTRIBUTING CO.
(Publishing Division)
Khushnuma Complex Basement
7, Meerabai Marg (Behind Jawahar Bhawan)
Lucknow-226 001, U.P. (INDIA)
Tel. : 91-522-2209542, 2209543, 2209544, 2209545
Fax : 0522-4045308
E-Mail : ibdco@airtelmail.in

First Edition 2008

ISBN 81-8189-266-6

Composed & Designed at :
Panacea Computers
Agarwal Sabha Bhawan
Subhash Mohal,
Sadar Lucknow-226 002
Tel : 0522-2483312, 94522 95008, 93359 27082
E-mail : prasgupt@rediffmail.com, prasgupt@hotmail.com

Printed at:
Salasar Imaging Systems
C-7/5, Lawrence Road Industrial Area
Delhi - 110 035
Tel. : 011-27185653, 9810064311

Preface

National Council of Devlomemnt Communication informaly came into existence in the year 1983. The first professional undertaking was publication of a Research Journal 'Interaction'. This Journal soon created space in the professional community internationaly. However, with the passage of time, it was felt to make it more focussed. Accordingly, from the year 2000, it started appearing as 'Journal of Communication Studies' (JCS). The Journal has completed 25 years of its publication and is widely distributed is the country and abroad.

A need to bring out edited volumes on the selected themes, was felt by the Editonal Board of the Journal. Agricultural Communication was in fact major component of publication, which still occupies pivotal place. The series of articles which are appearing in this volume, cover different facets of Agricultural Communication to be be classified as – impact of various media of communication on the acceptance of innovations, comparative analysis of various media, process of communication, fidelity, effectiveness of communication mesurment, models of diffusion, statistical tools of interpretation etc.

This publication will be an invalueable contribution academicaly specialy for the young researchers and budding scientists in the field of Social Sciences as a whole in the country as well as in the developing world. For those who are in the Northern Developed Nations, this will also provide an in-depth understanding of Indian Society. In the era of Information Revolution, in which we find ourselves, this publication will act as a Guiding Post, to indicate the direction, which we have covered and the way ahead. More so, it will fill the gap which we are witnessing presently in this field. Thus, it is a valuable coutribution from the Council and go a long way in enhancing the quality of research in Developing Society.

PROFESSOR V.K. DUBEY
Chief Editor
Journal of Communication Studies

Contents

Chapter 1

The Context of Agricultural and Development Communication

Alan Hancock

A quick, arbitrary and a rather thumbnail series of glimpses into development-oriented communication projects is being presented in this article. Some are traditional, some innovative and some go beyond the usual run of experience. These have been deliberately included.

Are there any common factors in these programmes? The most successful ones have certainly been well planned; they have worked from, or built up, a sound institutional base; they have been clear in their objectives; they have been well integrated into development work; they have stressed practical action programmes. But perhaps more than anything else, they have been conscious of their audiences, basing their activities not only on needs assessment but also on a close and personal relationship with their communities. This emphasis on the user is a crucial part of the modern view of communication in development.

CASE STUDIES OF COMMUNICATIONS MEDIA AND DEVELOPMENT

Botswana Laedza Batanani : Which means 'the Sun is already up, so come and work together' is a project in Botswana which is built upon popular theatre and folk media. It began in 1974, designed to motivate people to help themselves and to take full advantage of development opportunities. Planned by local leaders and development workers, the 'community awakening' campaign takes the form of an annual series of village festivals.

Laedza batanani started out as an experiment in combating local apathy in an area which was plagued by scattered settlement patterns. Each festival is run by a team, as an integrated non-formal education project involving every kind of extension workers in the area, either as a local organizer or as a member of a mobile team of actors and animators.

Each year the campaign team tours the major villages and puts on a ninety-minute performance of drama, puppetry, dancing, singing and drumbeat poetry. During this performance, a number of problems woven into the drama are presented. Afterwards, the audience gets divided into groups to discuss what they have seen. The drama is based on improvisation, with a good deal of audience participation; so there is ample scope for introducing problem, issues and messages. Just recently, an innovation has been tried, following up the festival with a post-campaign programme of demonstrations, exhibitions and projects.

MALI

In the context of agriculture, the *Mali Livestock 11 project* got under way in early 1977, with the support of USAID. It is a fairly traditional project, using a mixture of radio cassette tapes, audio-visual materials and interpersonal channels for concrete agricultural objectives, introducing conservation techniques and range-management practices, improved breeding and production techniques, in order to raise nutritional and economic status and generate foreign exchange. Communication was seen as a support to the main agricultural thrust, which itself operated on three fronts. It had one programme in the Dilly region to develop, test and apply new techniques for dealing with dry lands and livestock problems; it had another programme and facilities in the principal city of Bamako for training extension workers. In the southernmost region, a 'new lands' programme concentrated on opening up new territory, focusing particularly on tset-se fly control.

The responsibilities of the communication unit were therefore quite traditional the production of audio-visual materials for teaching and extension work; the training of extension workers in communication methods and some limited evaluation programmes. The main interests of the project, so far, however (it has not yet been formally evaluated), is that these functions and roles have been found insufficient, because many social and cultural difficulties have been encountered which need communication support. On the one hand, the project has revealed very diverse patterns among the farmers, some of whom are migratory, others not; some in mixed economics, others not. Very little can be taken for granted on exposure to media and the audio-visual support programme has had to be taken very slowly, with more emphasis on traditional media, and on simple, realistic graphics. The communication channels are being used as much for audience, needs surveys and data collection as they are in direct programme support.

GUATEMALA

Another simple project initiated in 1973 is the *Basic Village Education* programme for peasant farmers in *Guatemala.* It uses low cost materials, mostly graphics, forums and interpersonal communication, to try to change farming practices and increase production. It endeavours, as an additional goal, to increase the effectiveness of extension workers, especially for their problem solving abilities. The project began by working with poor Spanish speaking farmers in the south-east, but eighteen months later, it added radio to the media mix and turned to a different group, this time to Indian fanners in the western highlands. Two local radio stations were set up especially for the project, broadcasting eight hours a day and mixing entertainment with agricultural programmes.

The project was experimental and the various communication mixes were separately evaluated, including combinations of extension work. The result, contrary to expectations, was that radio alone had by far the greatest impact. But this was a very careful radio design, using custom-built radio stations; the role of the extension workers was crucial. Altogether, different results have been obtained from other Latin American projects which rely only on existing commercial radio stations.

COLOMBIA

Moving further along the scale of media complexity into multimedia programmes, one of the best known communication projects of all time is ACPO (Accion Cultural Popular) in Colombia. Its aim is to provide subsistence to farmers with a basic education. It began in 1947 at the initiative of a 25 year old Catholic priest, Father Salcedo. It was broadcast from a single radio station, Sutatenza.

It has, over the years, become an institution. It broadcasts basic education courses in literacy, numeracy, health, hygiene, economics, personal development etc. It trains a staff of over 900 for communication and development work. It runs a weekly newspaper and operates a printing press. It produces correspondence material and offers short extension courses. It sells hundreds of thousands of books each year, at cost, to the *campasinos;* sends entertainment and education vans into the countryside. It has broadened its services into the family planning area through a progarmme of "responsible parenthood".

The core of the programme is still the radio station, which ambraces 22,000 study groups, all organized by local campesions, who are given a basic training in group leadership and who then arrange meetings and keep basic records. These learning cells rely entirely on simple texts, especially on six, cost-free course text books which back up the broadcasts. It now reaches 140,000 campesinos, and more than 11,000 community organizers have passed through its training programmes. At least, 15 other Spanish speaking countries have modelled similar services on ACPO. It was evaluated recently, revealing a 23 per cent participation rate by villagers in ACPO programmes, in a representative sample of five communities, but the main indicator of its success is its survival.

SENEGAL

Another long-lasting project, which has also been institutionalized,is in *Senegal* in West Africa. This began as an initiative by UNESCO, with some extra support from the French and Canadian Governments, but since 1973, it has been totally supported by the Senegalese Government.

Its target audience was (and still is) Senegalese farmers, livestock producers and fishermen, amounting to roughly two million people. The aim is to equip the food producers with practical information, and especially with the opportunity to express their opinions regularly and effectively. It has used radio, film, correspondence and interpersonal channels.

The service is called *Radio Educative,* and it has always been built around radio programming. Initially, many administrative bottlenecks were experienced in getting the service started, so much so, that the President finally intervened (in 1968) and cut through the red tape. When the programmes finally got under way, they were based on local topics, like producing and marketing groundnuts, the responsiveness of government agencies to the people's and to farmer's needs, the ways in which debts are incurred and repaid in the villages etc. etc. At that time, 57 listening group leaders, or animators, were specially trained in leadership techniques. But the project placed a lot of emphasis on feedback through letters and recordings made in the field. Most of the impact was actually felt through this component, largely because the complaints that were made were broadcast, and then secured official recognition. It was largely as a result of this feature, for example, that groundnut prices were standardised, and, a number of peasant debts were cleared. By now, the listening groups have disappeared, but the dialogue with the radio station is said to be as alive as ever.

TANZANIA

In recent years, a number of communication projects have concentrated on community organization and self-development. Characteristically, these use the newer electronic media, but mostly in their smaller, group-based forms, like audio-cassette tapes and small-gauge video.

One project which used audio cassettes and listening groups to encourage women to participate in development programmes was based in *Tanzania*. This had two main objectives. The first was to involve women in planning, implementing and evaluating a development project which centered on needs experienced by the participants. The second was to use audio cassettes and other relatively inexpensive media, in a development education programme, and to test their effectiveness. Project components included a needs resource survey, conducted by the women participants, local production of audio tapes, discussions and planning by women and implementation of some of the plans. Two villages were selected as project sites and for experimental purposes also there were two control villages. Five group leaders from the two experimental villages were selected and given training in leadership and group dynamics. They, then virtually conducted the project.

Typical problems which came up were lack of water, poor sanitation, lack of clothing, poor nutrition and chronic drunkenness. They were identified during interview sessions and then tapes both, giving information and posing problems were produced by the leaders, in some cases by health educators. These were then used in group listening forums and some action programmes were set up. The research conducted showed that the audio cassette technology was workable, simple enough for participants to use; it also helped focusing attention on the discussions, and made the extension work more structured.

NEWFOUNDLAND, CANADA

Even more attention has been focused on video work, especially for community development. The *Fogo Island* experiments of the

National Film Board of Canada and the Memorial University of Newfoundland are probably the best known, and are among the oldest, experiments in this area. Originally conceived as a film project, they switched over to video as video technology improved and the possibility of instant replay gave the experiment a new orientation. Video was used, not to produce documentaries or instructional films, but to expose one community to another, and to open up debate within single communities. People saw themselves, and their convictions, replayed, juxtaposed with the sentiments of their neighbours. Most of all this, series of experiments (which has since been institutionalised) threw new light on the role of the development worker, in this case not so much a group leader as an *animator.* It is his role to provide instruments for social dialogue, without directly guiding or intervening; he has a catalytic function.

ALASKA, U.S.A.

But video does not exhaust the range of communications media in development. Moving to a different level of technology altogether, the satellite, an interesting example comes from *Alaska,* where the same experimental satellite which allowed the Indian SITE to take place, also made possible an experiment in telemedicine in 17 villages, in the Tanana region of the far north of the U.S.A. Initially, this used voice-only communication by satellite, employing modified taxicab radios and similar antennae to those used in SITE. By providing reliable communication channels between the villages and the district hospital, a medical support system was created and this meant that village health aides could provide health care services under a doctor's supervision. These aides were nominated by their communities and were trained for sixteen weeks by the Public Health Service, both in the field hospital and on the job. Each of them left the course, equipped with a drug kit, a diagnostic reference manual, instruments and of course new knowledge. At a scheduled time each day, the aides would take a 'doctor call' from the local hospital and receive medical advice on village patients.

A further experiment in 1974 used the ATS-6 satellite to add a video component, with visual, two-way links and computerised health records. This permitted diagnosis by doctors of more difficult village cases and also a referral service for patients. However, once the ATS-6 satellite was withdrawn, the service returned to its simpler form, and this basic service produced a 400 per cent increase in contacts between doctors and patients over a five year experimental period.

INDONESIA

Yet another new programme, aimed to link the widely separated campuses of Indonesia's Eastern Islands University Association (BKS), spread over more than 1600 miles of ocean, through satellite technology, with an emphasis on simple applications, especially on audio teleconferencing. This should mean that scarce specialised professors, research findings and library resources will be made more widely available to the rapidly increasing numbers of students throughout the region. The emphasis of the project was on improving the delivery of agricultural education and it involves a host of agencies and institutions such as the Ministry of Education and Culture, the Universities themselves, the Ministry of Communications, PERUMTEL, an organisation which owns and manages Indonesia's PALAPA satellite. Technical assistance is provided through the USAID programme. This project established a satellite conferencing link for eight universities, an extension campus, and the Bogor Agricultural Institute (covering altogether 350,000 square miles and more than 14 million people). At each site there was a room, large enough for fifty people, equipped with a speaker system, microphones and audiographic equipment such as a telecopier, electronic blackboard, or slow scan video for sending simple audio-visual information. The service was completely interactive and a portable earth station was used to extend the conferencing network to one or two remote extension centers.

It appears that modern views and theories of development have five main characteristics:

(i) They stress *self-reliance,* both at the individual and at the collective levels.

(ii) They are *user oriented;* the development process begins from an assessment of needs, not from the research laboratory.

(iii) They are *integrated;* they are not rooted in a particular economic or administrative sector but stress comprehensive development across all sectors.

(iv) They are *participatory;* stressing that as far as possible the clients of the development process should be involved in establishing the goals of development as well as its outcomes and be drawn into policy and planning as well as the implementation.

(v) They emphasise individual fulfilment and *quality of life.*

Given this focus, communication assumes a rather different role than hitherto assumed. It is no longer development *support* communications. Indeed, that term has been dropped from many vocabularies. It recognizes that development is an ongoing process with no single model and assigns communication a part to play at all stages, including diagnosis.

COMMUNICATION PLANNING

Consequently, let us say something about the planning process itself, because many of the newer concepts of communication within development really emerge from a policy and planning perspective.

Figure-1 includes, on the left hand side, a characteristic planning sequence, beginning with policy making, going on to strategic planning and continuing through implementation to regeneration or remodeling.

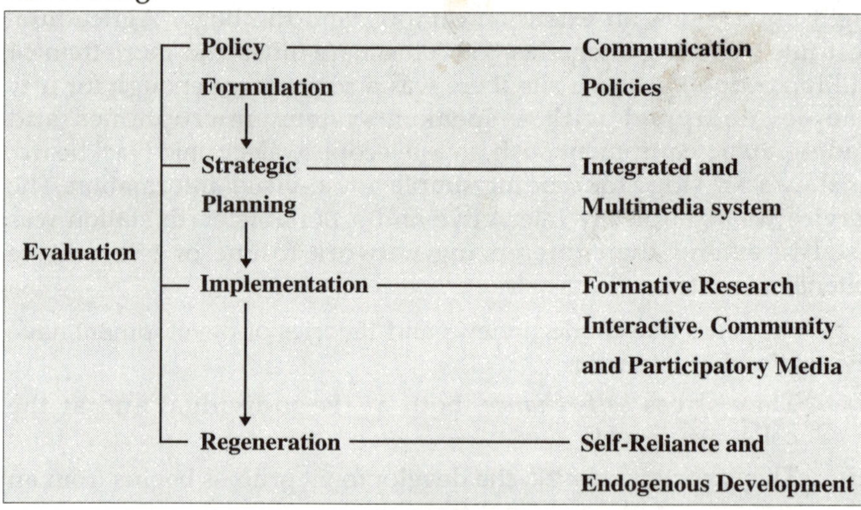

Fig. 1

This sequence (which is cyclic and interactive, rather than linear) is common to all programmes, whether, they are pitched at the level of a national campaign, or a local community initiative. So on the right hand side of Figure-1; one or two concepts have been located which are now important for development communication work, as they relate to the broad components of the planning cycle.

COMMUNICATION POLICIES

There is, first of all, the concept of *communication policy.* You may have read of the series of intergovernmental conferences on communication policies convened by UNESCO; the conference for Asia and Oceania took place in Malaysia in 1979. Communication policies are defined as "sets of principles and norms established to guide the behaviour of communication systems"; this means in practice that if communication media and processes are to be significant within the development process, then they must be structured and planned in such a way that they can undertake this role.

Communication policies are particularly concerned with the rationale of communication use. Consequently, they reflect the view of a society or agency about the development process. For example, a major writer on Latin American problems in recent years has been Paolo Freire, a Brazilian educator who, in his book THE PEDAGOGY OF THE OPPRESSED, held that the main constraint upon development was the indifference of the peasant, the apathy generated by an impossible set of socio-economic conditions, and that what was needed was 'conscientisation', or the creation within the peasant's own psyche of a sense of urgency, of aspiration, of drive and involvement, without which, he argued, the traditional diffusion models had no relevance at all in the Latin American case.

It is not that this model of conscientisation is universally relevant. But it represents the kind of attitude to the development process which needs to be articulated, in order to frame a coherent communication policy. The Freire model which is one of radical change and transformation has been translated practically into an emphasis on alternative approaches to the use of media. It favours smaller forms, such as video, which are not dependent upon a national, state-controlled or commercial media apparatus.

Policies also embrace attitudes to planning and management. The traditional development model is mirrored by centralized planning, top-down communication, and these are also embodied in media organisation. The participatory model suggests decentralised planning, a good deal of regional and community autonomy, leading in the end to so-called 'open' planning, in which the devolution of authority in extreme. Translated into operational terms, this may suggest forms of self-management for media organisations, such as have been developed in Yugoslavia.

COMMUNICATION STRATEGIES

The most significant new concepts in communication and development are found in the policy area, which has been neglected in the past, but other new concepts also cluster around strategic planning, mainly because their emphasis is on communication as a process rather than a set of instruments or techniques. And this implies in turn a kind of team approach, in which communicators work with other development planners from the earliest possible stage, to determine together what kind of development outcome is sought, and how communication may, among other agents, help to achieve it. The emphasis is upon integrated and cross-sectoral planning.

IMPLEMENTATION

There are also a number of new approaches at the level of implementation. A most important growth, which has been (in developing countries) particularly closely associated with India, and especially with work at Ahmedabad, as part of the strategy for the SITE experiment, is in the field of formative evaluation and formative production research. According to this approach, production of programmes and materials is based, first of all, on needs assessment of the clients the audience using social science and anthropological techniques to find out priorities and preferences. This is a difficult procedure, as people who are not used to modern media like television have no reference frame, and so their feelings can only be ascertained by preparing experimental productions and gauging reactions. An extension of this needs assessment base is in the use of prototyping, in which, before whole series of programmes are developed, pilots are made and tried out in limited, often partially controlled conditions, to see if they actually succeed in achieving the goals which they have set for themselves. The approach has two concrete advantages: it makes the final materials more relevant, more focused, and more acceptable; and it saves time and resources, by making it possible to correct imperfections at an early stage before major resources of production and distribution have been committed.

It is at the implementation stage too, that the kinds of local approaches which stress interactive and participatory communication are mainly found. There are difficulties and dangers in such programmes. They may antagonise and run foul of local bureaucracies and so be abruptly terminated. There has been some work of this kind in Asia, including India. It lies, more at the core of community media work in the industrialised countries, where it is being used to help break the stranglehold of highly organised mass media, especially television.

REGENERATION

In the end, let us discuss regeneration too. In applying communication media to development problems, we are engaged in a process of technology transfer,. in which external instruments and techniques especially for the modern media are being borrowed, in order to be adapted to new purposes. In many ways, this is similar to the kind of technology transfer with which we are all familiar, for example, the transfer of agricultural technology. There is the same need to consider technology as something more than a set of techniques and processes; to see that it embraces the infrastructures through which technology operates, the body of knowledge which is being transferred, the means of education, training and dissemination employed. But in communication technology transfer, we are dealing with the transmission of messages and cultural content, which complicates the process. We may say that the technology transfer process has four main components (as expressed in Figure-2): namely, Acquisition, Absorption, Adaptation and Regeneration. The knowledge to be employed is first sought out (or in the case of most transfers, pressed upon the recipient); ideally, it is then reviewed, understood and absorbed, before being adapted to a particular context. The goal of this process is not simply mastery or assimilation, important though these are, it is rather the ability to use the knowledge gained to begin from again scratch and develop new, appropriate technologies.

Technology Acquisition

↓

Technology Absorption

↓

Technology Adaptation

↓

Technology Regeneration

Fig. 2.

CONCLUSION

This has been, quite deliberately, an introductory and in many ways simplistic paper, intended to illuminate new themes and concepts and their interrelation, but not to take the argument very much further. Its main argument, however, can be summarised in a very few lines. It is simply that agricultural communication cannot be seen in isolation; only as a part of development communication. In the same way, communications media cannot be viewed independently; they are subordinated to a broader view of overall development policy and strategy. And finally, agricultural communication will gain, if it is seen within a continuance of communication planning and technology assessment, in which the needs of the user, the rural audience, are the dominant concerns.

Chapter 2

A Paradigm of Attributes of Elements of Communication Process Contributing to Communication Fidelity

S. Pathak and A. K. Majumdar

Historically, if the approaches to communication and diffusion researches are scanned over the years, it would appear that there has been a change in the understanding of social reality. Initially, there was more dependence on polar types of social relationships like modern-traditional type of concepts of societies. Gradually, there was a change in the perception of social reality. Naturally, sociological perspective also changed. Our farmers are considered to be rational still maintaining 'familism' which is definitely not detrimental to the process of development. Therefore, the physical situations of the farmers (receivers) which has been considered here as agricultural infrastructure have some relevance in communication research. When the communication is considered in the perspective of development, there must be a comprehensive approach to communication i.e. it must be considered in the entire background of social reality.

Communication is a process which involves elements such as communicator, message, channel and receiver. Research studies have, however, indicated that there are certain characteristics pertaining to these elements of communication process which determine the effectiveness (fidelity) of communication process (Hovland et. al., 1953; Berlo, 1960; Leagans, 1961; Rogers, 1962; Verma 1971; Rogers and Shoemaker, 1971; Mortensen, 1972). Different characteristics are associated with different elements. The basic aim of communication is the production of desirable changes in the behaviour of receiver in terms of knowledge, skill, attitude and action. In other words, the change in the behaviour of receiver remains the sole objective of the communication.

The concept of fidelity which is very much important for the entire communication process has been studied sporadically and only piecemeal approach has been made by various social scientists during the past years in India. The effectiveness of communication has been studied mostly with reference to the adoption of an agricultural innovation. Very few studies have attempted to delineate the entire communication process and bring out the factors influencing the effectiveness of communication. Jha (1968), however, conducted a comprehensive study which took some of the evidences of behavioural change into a composite measure to determine the communication fidelity. The attributes pertaining to each element of communication process which might have some bearing on the communication fidelity have perhaps not been considered in a single study. Keeping this in view, this study was conducted to identify some important attributes pertaining to the communicator, message, channel and receiver which might determine the effectiveness of communication process. This paper deals with the contribution of attributes of elements of communication process in explaining variability in communication fidelity with reference to the improved practices of jute cultivation.

Methodology

The study was conducted in the Community Development Block Amdanga in the district of 24-Parganas (North), West Bengal. Two villages in each of 11 operational areas were randomly selected and 15 farmers from both the villages were then randomly selected from a list of the farmers. The total sample, therefore, consisted of 165 farmers. The respondents were personally interviewed for the collection of field data. The responses were recorded on five points rating scales. The scoring was done by assigning scores as 5-strongly agree, 4-agree, 3-undecided, 2-disagree and 1-strongly disagree. The aggregate score earned by an individual respondent on various items was taken as the score for an attribute.

Attributes of Elements of Communication Process

Attributes are considered as the characteristic or qualities possessed by or associated with communicator, message, channel and receiver. Attributes were selected and measured with the help of various scales, each comprising a number of items or statements initially selected on the basis of critical ratios. Attributes were finally selected with the help of agreement test. The method employed was that of Spearman's Rank Difference Correlation (Rho). The 'rho' coefficients were calculated on the basis of average scores earned by 30 extension personnel (communicators) and 44 farmers (receivers). The degree of agreement was determined between the ranks of items of different attributes in the two groups. The reliability of the scales was also tested by test-retest method through correlation analysis.

Communicator is the person who communicates the messages of the improved practices of jute cultivation to the farmers. The communicator may be extension personnel like Block Development Officer, Agricultural Extension Officer, Village Level Worker, Ex tension Assistant, Technical Assistant, Agricultural Scientist, a village leader or group of them. Attributes are the characteristics or qualities of a communicator pertaining to his communication skill, knowledge, and attitude as perceived by receivers. Among 21 selected attributes of communicator, 7 belonged to his communication skill, 6 related to his knowledge and the other 8 covered his attitude.

Message is the improved practices of jute cultivation recommended and communicated to the farmers. In all, 13 attributes of message as perceived by receivers were selected for this study.

Channel refers to the extension method through which messages of the improved practices of jute cultivation were communicated to the farmers (receivers). The channel may be of the type of personal contact, group contact or mass contact or all of them. Eleven attributes of channel as perceived by the receivers were selected for this study.

Receiver of the message of the improved practices of jute cultivated is the farmer- the actual user. Attributes are the characteristics of a receiver as perceived by himself. Among 18 selected attributes of receiver, five belonged to his communication skill, five related to his knowledge and eight covered his attitude. In addition, seven background variables were also included in the study.

Details of the selected attributes pertaining to the communicator, message, channel and receiver are given in Table 1.

Table-1 : Selected Attributes of Communicator, Message, Channel and Receiver with their Critical Ratios, Rho-coefficients and Correlation Coefficients

S. No.	Attributes	No. of items selec.	Critical Min.	Ratios Max.	Rho-coefficients	Correlation coefficients
	Communicator					
1.	Encoding and decoding skills	5	2.30*	5.03**	0.93*	
2.	Presentation skills	4	2.10*	4.28*	1.00*	
3.	Technical skill	4	2.20*	2.85*	1.00*	
4.	Skill in selection and use of communication materials	4	2.20*	2.85*	1.00	0.93**
5.	Change-effecting skill	5	2.33*	5.60**	0.94*	
6.	Motivating skill	5	2.33*	5.60*	0.94*	
7.	Dealing skill	4	2.70*	4.17*	1.00*	
8.	Knowledge of message	6	2.12*	5.59**	0.84*	
9.	Knowledge of channel	6	3.08**	4.20**	0.85*	0.89**
10.	Knowledge of receiver	8	2.32*	5.21**	0.76*	
11.	Knowledge of self skills and attitude	5	2.68*	5.19**	0.93*	
12.	Knowledge of communication techniques	6	2.29*	4.48**	0.87*	0.89**
13.	Knowledge of communication background	5	2.20*	3.71**	0.96*	
14.	Attitude towards self general behaviour	9	2.35*	5.51**	0.82*	
15.	Attitude towards self knowledge	5	4.28**	5.98**	0.91*	0.90**
16.	Attitude towards accuracy of self thought	7	2.10*	6.51**	0.72*	
17.	Attitude towards self reading habit	4	2.10*	3.88*	1.00*	
18.	Attitude towards self writing habit	4	2.15*	2.77*	1.00*	
19.	Attitude towards self speaking habit	4	2.15*	3.62**	0.92*	
20.	Attitude towards message	10	2.31*	7.97**	0.79*	0.90**
21.	Attitude towards receiver	15	2.16*	6.08**	0.63*	0.92**
	Message					
1.	Attractability	5	2.63*	4.20**	0.91*	
2.	Motivability	6	2.13*	2.92**	0.86*	

S. No.	Attributes	No. of items selec.	Critical Min.	Ratios Max.	Rho-coefficients	Correlation coefficients
3.	Understandability	4	2.99*	5.60**	1.00*	0.93**
4.	Informativeness	4	2.77*	5.63*	1.00*	
5.	Probifability	5	2.36*	2.95**	0.94*	
6.	Practicability	5	2.53*	6.39**	0.94*	
7.	Attainability	7		2.92**	4.51**	0.73*
8.	Need-compatibility	6	2.35*	2.92*	0.83*	
9.	Accuracy and adequacy	5	2.53*	5.85**	0.91*	0.93**
10.	Suitability	4	2.83*	4.79**	1.00*	
11.	Cultural-compatibility	5	2.19-	3.59*	0.95*	
12.	Economy in time and labour	8	2.39*	2.98**	0.89*	
13.	Receivability	4	2.81*	5.19**	1.00*	
	Channel					
1.	Attractability/attention getting	5	2.97**	6.71**	0.92*	
2.	Motivability	6	2.12*	7.79**	0.83*	
3.	Educatability	5	2.20*	6.71*	0.96*	
4.	Dependability	4	2.54*	5.60**	1.00*	
5.	Suitability	8	3.46**	7.09**	0.69*	
6.	Effectiveness	5	2.10*	6.39**	0.90*	
7.	Simplicity	5	3.52**	7.79*	0.94*	0.89**
8.	Fast-approachability	5	2.78*	4.64**	0.93*	
9.	Projection potentiality	6	2.60*	4.89**	0.90*	
10.	Impact universality	5	2.17*	6.04**	0.92*	
11.	Social-contactability	5	2.19	6.99*	0.95*	
	Receiver					
1.	Message understanding and interpretation skill	7	2.85**	5.85**	0.75*	0.96**
2.	Knowledge acquiring skill	6	3.04**	4.79**	0.83*	
3.	Technical skill	7	2.17*	5.94**	0.86*	
4.	Message utilization skill	6	3.04**	7.15**	0.87*	0.96**
5.	Problem analyzing skill	6	2.15*	6.71**	0.84*	
6.	Knowledge message	8	2.38*	4.79**	0.68*	
7.	Knowledge channel	5	2.50*	7.77**	0.91*	
8.	Knowledge of communicator	11	3.33**	6.95**	0.79**	
9.	Knowledge of self skill and attitude	7	3.41*	5.36**	0.93**	0.94**
10.	Knowledge of communication background	4	2.36*	6.72**	1.00*	

S. No.	Attributes	No. of items selec.	Critical Min.	Ratios Max.	Rho-coeffi-cients	Corre-lation coeffi-cients
11.	Attitude towards self general behaviour	12	2.10*	5.24**	0.60*	
12.	Attitude towards self knowledge and alertness	5	2.24*	9.57**	0.90*	
13.	Attitude towards self social adjustment	6	2.86**	7.79**	0.92*	0.97**
14.	Attitude towards self straight thinking	4	3.90**	9.92**	1.00*	
15.	Attitude towards self listening habit	5	2.71*	4.97**	0.95*	
16.	Attitude towards self reading habit	5	3.71**	15.50**	0.90*	
17.	Attitude towards message	15	2.47*	9.27**	0.60**	0.98**
18.	Attitude towards communicator	28	2.77*	11.21**	0.57**	0.96**

*Significant at 0.5 level

**Significant at .01 level

Communication Fidelity :

Communication is the process of the flow of the messages of improved practices of jute cultivation from extension personnel and agricultural scientists to the farmers. Fidelity is the faithful performance of the communication process by all its elements i.e. communicator, message, channel and receiver. The faithful performance means the effectiveness of communication. The communication fidelity is therefore, the positive effect of communication on farmers' behaviour in terms of amount of knowledge gained, kind of attitude formed and extent of adoption of some of the improved practices of jute cultivation, each for a unit of communication sensitivity at the time of this study. The introduction of communication sensitivity in the measurement of fidelity was desirable because a measure of fidelity must take into account the potential of the communication act itself which might vary from farmer to farmer depending upon the mutual interplay of all the communication elements varying from situation to situation. Fidelity is, therefore, a composite measure taking into account all the above components and was measured according to the method developed by Singh and Jha (1970) in the form of a ratio as :

$$F.I. = \frac{F_1}{F_2} \times 100 = \frac{\frac{K_1}{S_1} + \frac{A_1}{S_1} + \frac{P_1}{S_1}}{\frac{K_2}{S_2} + \frac{A_2}{S_2} + \frac{P_2}{S_2}} \times 100$$

Where F.I. = Fidelity Index

F_1 = Fidelity score obtained by an individual

$$\text{Receiver of message} = \frac{K_1}{S_1} + \frac{A_1}{S_1} + \frac{P_1}{S_1}$$

(K_1, A_1, P_1, and S_1 being knowledge, attitude, adoption and communication of an individual receiver of message.)

F_2 = Possible Fidelity score that could be obtained by an individual receiver of message when K, A, P, and S are maximum possible =

$$= \frac{K_1}{S_1} + \frac{A_1}{S_1} + \frac{P_1}{S_1}$$

(K_2, A_2, P_2, and S_2 being maximum possible knowledge, attitude, adoption and communication sensitivity score respectively that an individual receiver of message could secure).

Results and Discussion

The technique of multiple regression analysis employed to determine the contribution of the attributes of each element of the communication process to the communication fidelity. Accordingly, four regression analyses, one for each element, were computed. The results of regression analyses with significant partial regression coefficients are presented in Table 2 and are discussed as follows :

1. Contribution of attributes of communicator to communication fidelity

Out of 21 communicator's attributes put together in regression analysis, the partial regression coefficients were positively significant in the case of 12 attributes such as presentation skill, motivating skill, dealing skill, knowledge of channel, knowledge of communication techniques, knowledge of communication background, attitude towards self knowledge and alertness, attitude towards the accuracy of self thought, attitude towards self reading habit, attitude towards self writing habit, attitude towards self speaking habit and attitude towards the message. These attributes contributed significantly to the prediction of communication fidelity of the respondents with regard to the improved practices of jute cultivation. Similarly, teaching ability, communication skill, effective speaking and writing and knowledge were reported to be crucial factors for the effectiveness of village level workers communicators (Dubey, 1958; Berlo, 1960; Kelsey and Hearne, 1967) in the communication fidelity. This finding concurs with the propositions earlier propounded by Berlo (1960) and Leagans (1961) that a message should be clear by understandable by the audience, specific and accurate supported by facutal materials, economically and socially suitable to the needs, interests and values of the audience (receivers). Jha (1968) pointed out that among various factors, profitablity of the message contributed to the variability in the communication fidelity. Pandey and Roy (1978) reported that message related to felt needs and attention getting were two important factors for its effectiveness.

3. Contribution of attributes of channel to communication fidelity

The multiple regression analysis revealed that the partial regression coefficients were positively significant in the case of attractability, suitability, simplicity, impact-universality and social-contactability of the channel. These attributes contributed significantly to the prediction of communication fidelity of the respondents. Wittich and Schuller (1959) and Hindlekar et. al. (1970) reported that simple and attractive illustrations were highly effective in communicating ideas.

The R^2, however, reveals that 58.86 percent of the variation in the communication fidelity of the respondents could be explained with the help of 11 attributes of channel. Jha (1968) found six crucial factors including frequency of the use of the channel which contributed to the variability in communication fidelity.

4. Contribution of receiver's attributes and background to communication fidelity

Six variables namely receiver's message understanding and interpretation skill, knowledge acquiring skill, attitude towards self general behaviour, attitude towards communicator, psychological makeup and agricultural infrastructure significantly and positively

Table-2. Partial regression coefficients of significant attributes of elements of communication process with communication fidelity

S. No.	Attributes of elements	Partial regression coefficients	't' value	
	Communicator			
1.	Presentation skill	2.3625	2.62292**	
2.	Motivating skill	1.5637	2.5953**	
3.	Dealing skills	1.1850	2.1899*	
4.	Knowledge of channel	0.8235	1.9916*	
5.	Knowledge of communication techniques	0.9436	2.0936	R2 = 0.6354** (with 21 independent variables)
6.	Knowledge of communication background	0.9013	1.9969*	
7.	Attitude towards self knowledge and alertness	0.9318	2.0550*	
8.	Attitude towards accuracy of self thought	0.8904	2.1857*	
9.	Attitude towards self reading habit	0.9028	1.9947*	
10.	Attitude towards self writing habit	3.1806	4.7471**	
11.	Attitude towards self speaking habit	0.9905	1.9832*	
12.	Attitude towards message	0.3967	1.9679*	
	Message			
1.	Understandability	1.6545	2.2252*	
2.	Accuracy and adequacy	0.9965	1.9784*	R2 = 0.5592** (with 13 independent variables)
3.	Suitability	1.7088	2.1444*	
4.	Receivability	1.2873	1.9948*	
	Channel			
1.	Attractability	1.0602	1.9827*	
2.	Suitability	0.5522	1.9922*	R2 = 0.5886** (with 11 independent variables)
3.	Simplicity	0.7530	2.1060*	
4.	Impact-universality	1.0858	2.0326*	
5.	Social-contactability	1.3796	1.9925*	
	Receiver			
1.	Message understanding and interpretation skill	0.5676	2.0558*	
2.	Knowledge acquiring skill	0.7815	2.0490*	
3.	Attitude towards self general behaviour	0.2950	2.0178*	R2 = 0.6441** (with 25 independent variables)
4.	Attitude towards communicator	0.2547	3.3139	
5.	Psychological makeup	0.0954	2.1689*	
6.	Agricultural infrastructure	0.0659	2.0558*	

*Significant at .05 level,

**Significant .01 level

contributed to the prediction of communication fidelity of the respondents. Similarly, Singh, 3ha (1970) reported that communication fidelity depended significantly and positively on six independent factors such as communicator's language compatibility with that of receiver, communicator's degree of contact with receiver, profitability of message, frequency of the use of channel, socio-economic status of receiver and receiver's past experience with communicator's message. In another communication, Pathak and Majumdar (1981) reported that three attributes of communication elements namely receiver's attitude towards communicator, psychological make-up and agricultural infrastructure positively and significantly influenced the communication fidelity of the respondents.

The R shows that 25 variables accounted for 64.41 percent of the variation in communication fidelity. Singh and 3ha (1970) reported that six variables as mentioned in preceding paragraph accounted for 99.20 per cent of the variability in communication fidelity. Pathak and Majumdar (1981) found that seventeen attributes of communication elements were responsible for 79.79 percent of the variation in communication fidelity of the respondents.

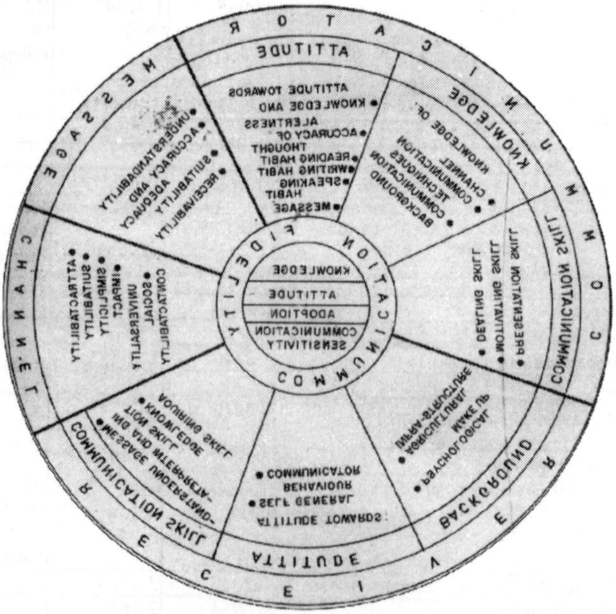

A paradigm of attributes pertaining to different elements of communication process contributing to communication fidelity

From the foregoing results, it is evident that there are 27 attributes of elements of communication process which have significantly and positively contributed to the variability in communication fidelity of the respondents with reference to the improved practices of jute cultivation. All the 27 significant attributes are paradigmatically presented in Figure-1. The paradigm also illustrates the components of communication fidelity in terms of behavioural changes. The paradigm will help in identifying various important attributes pertaining to different elements and their inter-relationships.

A few implications of this study are noteworthy. The result supports the earlier studies that the effectiveness of a communicator depends on his communication skill, knowledge level and attitude. This study made in depth attempt and brought into focus certain other characteristics (attributes) which affected communication fidelity. Three attributes constituting communication skill emerged significant were presentation skill, motivating skill and dealing skill. This suggests that in order to make communication more purposeful and effective, efforts should be made to improve communication skill of extension personnel. In interpersonal communication which is the basic theme of delivering scientific know how to the farmers in India, emphasis should be given on methodological training on how to introduce messages and present them more effectively to the receivers, how to stimulate and arouse the interest and desire of receivers and motivate them for adoption of agricultural innovations and how to deal and approach them tactfully and democratically. A change agent with technical knowledge and developed communication skill can effectively arouse the interest of farmers and motivate them for action.

Communicator's knowledge of various channels, knowledge of techniques necessary to communicate messages effectively and knowledge of communication background i.e. physical facilities available for communication appeared significant contributing factors affecting communication fidelity.

Six attributes constituting the attitude of a communicator emerged significant were: communicator's attitude towards self knowledge and alertness; attitude towards accuracy of self thought; attitude towards self reading habit; attitude towards self writing habit; attitude towards self speaking habit and attitude towards message. This demands that continuous programme of upgrading the quality and competence of the extension personnel should be undertaken for simultaneous development of their communication skill, knowledge and attitude keeping with the pace of fast changing agricultural technologies.

The findings of this study also reveal that four attributes of message namely understandability,accuracy and adequacy, suitability and receivability contributed significantly to the communication fidelity. This points out that the communication fidelity would be more when the extension personnel communicate messages which are complete, reliable and accurate; suitable and appropriate; clear and concise in understanding and could be sensed/received by the receivers (farmers).

Five attributes of channel i.e. attractability, suitability, simplicity, impact-universality and social-contractibility were significant contributing factors to the communication fidelity. It is, therefore, imperative that due consideration should be given to the appropriate channels which attract and hold the attention of the receivers, are simple in handling and use; are suitable in the existing situation; can develop personal acquaintance and social contacts and have universal appeal or impact on all types of receivers.

Four attributes of receiver namely message understanding and interpretation skill. knowledge acquiring skill, attitude towards self general behaviour, attitude towards communicator contributed significantly to the communication fidelity. This suggests that in order to make communication effective for achieving good results, the extension personnel (communicators) should help the receivers in acquiring useful messages and developing their skills in understanding and interpreting messages as accurately as possible for practical uses. What the farmers (receivers) expect is the imparting of the skills for using various messages/ideas about agricultural innovations to get maximum possible results. Merely acquisition of knowledge would not help the receivers (farmers) unless they understand and interpret messages as accurately as intended for. Efforts should also be made to influence the attitude of receiver towards his general behaviour, and create good impression of communicators (extension).

Two background variables i.e. psychological make-up of receiver and agricultural infrastructure facilities manifested their significant contribution to the communication fidelity. Psychological orientations of the farmers (receivers) can be developed/ augmented by providing farm literature and organising appropriate training programmes primarily based on behavioural aspects. There is also need for institutional changes to develop adequate agricultural infrastructure in the villages.

The most striking feature of this study is that two background variables i.e. receiver's psychological make-up and agricultural infrastructure facilities which had significantly influenced his perception of the attributes of elements of communication process (Pathak and Majumdar, 1981) had also significantly influenced and contributed to the variation in the communication fidelity. This study thus proves that receiver's background influences his perception (of the attributes) which in turn affects the communication fidelity. Background on the other hand also influences communication fidelity directly. Unlike industrialized countries where communication is message-oriented and more technology-oriented, interpersonal communication is still most credible method in India, such relationships are therefore, important for effective communication in agricultural development.

REFERENCES

1. **Hello, D. K.,** "The Process of Communication: An Introduction to Theory and Practice." Holt, Rinehart and Winston, New York, Inc., 1960.
2. **Dubey, S. C.,** 'India's Changing Villages: Human Factors in Community Development'. Rout ledge and Kegan Paul Ltd., London, 1958.
3. **Hindlekar, W. M.; Mohanty, B. B. and Lokhande, M. R.,** 'A study on the design of posters for rural areas.' Ind. J. of Extn. Edu. 6(3&4): 51-57, 1970.
4. **Hovland, C.; Janis, I; Kelley, H.** 'Communication and Persuasion'. Yale Univ. Press, New Haven, Conn.1953.
5. Jha, P. N., 'A Critical Analysis of Factors Associated with Communication Fidelity with reference to High Yielding Varieties Programme' Ph.D. thesis (Unpub.) Indian Agril. Res. Instt., New Delhi, 1969.
6. **Kelsey, L. D. and Hearne, C. C.,** 'Cooperative Extension Work'. Comstook Publishing Associations, Ithaca, New York, 1967.
7. **Leagans, J. Paul** 'The Communication Process in Extension Education in Community Development'. Directorate of Extension, Min. of Food and Agril., Govt. of India, New Delhi, 1961.
8. **Mortensen, C. D.,** 'Communication : The Human Interaction'. McGraw Hill Book Co., New York, 1972.

9. **Pandey, S. N. and Roy, N. K.,** 'Factors associated with effectiveness of farm broadcast.' Ind. J. of Extn. Edu. 14 (1&2) 54-56, 1978.

10. **Pathak, S. and Majumdai, A. K.,** 'Elements of Communication process and communication fidelity : a relational analysis'. Ind. J. of Extn. Edu. 17(1&2) : 8-22, 1981.

11. **Rogers, E. M.,** 'Diffusion of Innovation.' Free Press, New York, 1962.

12. **Rogers, E. M. and Shoemaker, F. F.** 'Communication of Innovation - A Cross - Cultural Approach'. Free Press, New York, 1971.

13. **Singh, K. N. and Jha, P. N.,** 'A paradigm of factors contributing to communication fidelity with reference to High Yielding Varieties Programme', Ind. J. of Extrt. Edu. 6(3&4) : 1-11, 1970.

14. **Singh, K. N. and Jha P. N.,** 'A concept of communication sensitivity and its measurement with reference to High Yielding Varieties Programme.' Ind. J. of Soc. Res. 12(2), 1971.

15. **Veima, H. S.,** 'Personal influence in the transmission of ideas, in Communicating Agricultural innovations : A strategy for India. 'H. G. Rieger (Ed.) Navchetna Prakashan, Varanasi (U.P), India, 1971.

16. **Witlich. W. D. & Schuller, C. F.,** 'A.V. Materials-Their Nature and Use'. Harper & Brothers, New York, 1959.

Chapter **3**

Measurement of Communication Act Efficiency of Extension Personnel under Systems Perspective

M.M.P. Akhouri

A method to measure the Communication Act Efficiency under Systems Perspective has been developed and used in the study. The output indices which were used as a measure of effectiveness showed highly significant correlation coefficients separately with peer and administrative evaluated effectiveness. But the non-significant correlation coefficients of Communication Act Efficiency with those of peer and administrative evaluated effectiveness suggests the lack of appreciation of cost incurred and facilities availed by extension personnel for performing the act. This method of measuring Communication Act Efficiency helps categorizing extension personnel into different levels of efficiency and comparing one category of extension personnel with other group.

'A communication system is more or less efficient, given the degree of communication effectiveness, it permits or facilitates, relative to the cost involved in running it or maintaining it. There is inevitably some cost involved in generating, disseminating, acquiring and consuming information. This ratio of overall effectiveness of any communication system to the composite cost of generating, disseminating, acquiring and consuming information within it is one rough measure of its efficiency' (Thayer, 1968).

The process of modernizing developed society involves three distinct systems research system that creates new knowledge and innovations; the client system consisting of potential users of new knowledge; and the extension system comprising of extension personnel performing the task of communication link between the research and the client system. The quality and quantity of effective communication among the members, both within and between these systems, is the crux to modernization, be it in the field of agriculture, industry, health or any developmental area. Behavioural scientists have expressively studied the communication behaviour of members of client and research systems under two research traditions diffusion of innovations and scientific communication. The almost left out area relates to communication behaviour of members of extension system.

Studies related to extension personnel are few. To illustrate, Brown and Deekens (1958) attempted to study the role of extension specialists. Boone (1959) studied the professional status of extension specialist as compared with research and teaching staff. Similarly, Indian scholars like Dube (1958); Sengupta (1966); Rahudkar (1963); Salvi and Dudhani (1967); Singh and Prasad (1972); and Chakravorty (1971); studied some aspects like personal characteristics, role and job satisfaction of extension personnel, mostly the Village Level Workers. None of these studies touches the communication behaviour of extension personnel.

Particularly, the communication efficiency of extension personnel is rather almost untouched. Efforts are almost absent about formulation of methods and identification of dimensions to be included for measuring the communication efficiency. The present report describes an attempt to conceptualize, formulate and finally measure the communication efficiency of Extension personnel. This effort formed the part of a comprehensive study Communication Behaviour of Extension Personnel. An Analysis of Haryana Agricultural Extension System; conducted by the author (Akhouri 1973).

Locale of the Study

This study was planned in a comparatively new and small state, Haryana. The data were collected during 1972-73. The research design selected for the study was field survey in an organizational setting. The twin organizations of the state, Department of Agriculture and Directorate of Extension Education of Haryana Agriculture University, were defined in terms of one Haryana Agricultural Extension System under systems perspective.

Sampling

The sample for the study included a total of 166 extension personnel of the following sub-systems :

University level Extension Specialists (ES)	7
District level Extension Specialists (AES)	56
Zonal Officers (ZO)	7
Area Officers (AO)	15
Agricultural Inspectors (AI)	81
Total	166

In case of the first three sub-systems, entire population was included in the study. But for selecting Area Officers and Agricultural Inspectors from last two sub-systems; first 20% blocks from each district of the State were randomly sampled. All Area Officers and Agricultural Inspectors of these sampled blocks were included as respondents.

Tools and Techniques Used

Tools used for data collection from the selected extension personnel were personal interviews with the help of structured schedule consisting of a number of indices, office records reports, important official circulars and other departmental publications were further used as resource material.

Formulation of a Measurement

In view of non-availability or non-existence of studies of communication efficiency of Extension Systems or Extension personnel, the Thayer's "rough measures" referred earlier were used as base to develop a measurer for Communication Act Efficiency refers to communication effectiveness and composite cost as two crucial variables. Communication effectiveness, in many studies, on extension methods and communication media have been reported in terms of changes in farmers' behaviour. This approach can as well be adopted for measuring communication efficiency of communicators and extension workers. Alternatively, under systems perspective output of a communication system may be viewed as an effect of all communication activities performed by individual component of the system. Of course, the output will be the total effect of all communication activities like acquisition, processing and disseminating activities and can very well be regarded as effectiveness of communication Acts instead of effectiveness of Communicator. Accordingly, the efficiency measured by using the concept of communication-act-effectiveness as named as Communication-Act-Efficiency and not the communication efficiency.

Thus, communication-act-efficiency hereafter known as CAE, was defined as the degree of communication-act-effectiveness displayed by extension personnel relative to the cost involved; facilities for communication act available and proportion of time allocated to extension personnel for these communication acts. Based on the above definition, a mathematical formula was developed for the computation of CAE.

$$CAE = \frac{Oj}{(Ci+Fi)\,Jc}$$

Where,

Oi = the output of an extension worker expressed in terms of Output Index value,

Ci = is the cost of an extension worker expressed in terms of Cost Index value.

Fi = is the facility available with extension worker expressed in terms of Facility Index Value.

Jc = is the job commitment of an extension worker expressed in terms of percentage of time allocated for communication acts.

Information Output Behaviour

The information output behaviour refers to all such activities performed by extension personnel for dissemination of scientific and technical information related to their role. The dissemination of information may either be direct to the member of client system or through different levels of extension personnel. For example, an university level extension specialist, at one time, may be lecturing to agricultural inspectors and on anotheroccasion, he may be explaining about the package of practices for the farmers. In both cases, his activities are part of his output behaviour. Information output behaviour of extension personnel also includes their activities related to the dissemination of feedback information. The upward communication of farmers' need, problem and reactions regarding some innovation, forms the part of their information output behaviour.

The amount of information output made by an individual extension person was measured through an Information Output Index. This output index consisted of a number of items which are usually engaged by extension personnel for transmitting the information from their own system to other systems. These items and the procedure of scoring data against such items have been reported by Akhouri (1973). Each item had numerical values. Because these items had different ranges and units of measurement; the raw scores were converted to Z-scores. The Z-scores of all items were added up to denote the Information Output Index value of individual extension personnel.

Cost Index

In order to calculate efficiency, Thayer (1968) suggested to consider composite cost of generation, disseminating, acquiring and consuming the information. Costs on generating and consuming information were not involved in case of the Extension System, because information was generated in Research System and consumed in Client System. The Extension System was concerned only with acquisition, processing (which Thayer omitted), and dissemination of information. The cost involved on these three activities were broadly categorised under two categories;

(i) Cost in the form of pay and allowance of extension personnel and the member of their establishment, if any.

(ii) Cost of facilities such as typing, cyclostyling, telephone, transport, funds for correspondence etc.

Each of the two types of cost was measured by separate indices - Cost Index and Facility Index for first and second type, respectively.

For calculating Cost Index, data were obtained regarding the pay-scales of all extension personnel and also of the members of their establishment, if any. The following items were included in the cost of personnel;

1. Mid pay in the pay scale of the post of the incumbents and the permissible allowances on that amount.
2. Mid pay in the pay scales of the posts of members of establishments of particular extension workers and the permissible allowances thereon.
3. Traveling Allowances permissible to different extension personnel and members of their establishment, if any.
4. Amount of contingency sanctioned to each member of extension personnel.

In case of joint establishment and contingency, the individual's share was calculated by dividing the joint amount by the number of extension personnel belonging to the same establishment. After totaling the amount of all the above four items, cost score for individual member was obtained. The total amount was divided by 10,000 for reducing the score in small figure. The score was converted into standard scores.

Facility Index

Every communication act requires certain facilities. In formal extension organization these facilities may be typing, cyclostyling, photography, artist, slide projector, movie projection, telephone, printing etc. A list of 14 such items of communication facilities was prepared after reviewing the literature and concerned persons from field. Data regarding available facilities were obtained by asking one question from each extension person.

"How many of the following facilities are available to you for communication act?"

The items present for checking included, typing facilities, cyclostyling, photography, slides projection, movie projection, slides and film strips, movie films, exhibition unit, public address set, demonstration equipment, telephone, printing and paper and other writing materials. One score was given for each checked item. The total score obtained by each individual was converted into facility index score by transferring it to its standard score.

Job Commitment

Job commitment is defined as the degree to which an individual is committed to perform a given job. Through this, an attempt was made to measure the extent to which extension personnel spend their working time in performing communication acts. This was measured in terms of percentage by asking the following questions:

"Approximately what percentage of your time do you actually spend on the following activities?"

1. Extension work (including extension training),
2. Class room teaching,
3. Research work, and
4. Administration.

The percentage of time indicated against extension work was taken as job commitment of extension workers for communication acts.

Measuring Communication Act Efficiency Sub-system-wise

The average CAE of Extension Personnel of Haryana Agricultural Extension System comes to 2.48 with a standard deviation of 1.55. Sub-systemwise average CAE as presented, in Table 1 reveals that AI have maximum CAE followed by AO, ZO, AES and ES. Although, ZO have maximum facility and cost indices which are the devisor in the formula for calculating CAE; ZO emerge more efficient than ES and AES. However, the apparent differences between the average CAEs are not statistically significant except only in three cases- between AI and AES, Al and ES and also between AO and ES.

Table-1. Sub-system wise average CAE and significance of their differences

SD	Average	Sub-systems	Significance of t-value in respect to				
			AI	AO	ZO	AES	ES
1.03	3.08	AI	-				
1.29	2.89	AO	0	-			
0.56	1.94	ZO	0	0	-		
0.38	1.78	AES	*	0	0	-	*
1.03	1.74	ES	*	*	0	0	-

*Significant at .05 probability level and 0 Non-significant at .05 level of probability.

All extension personnel have been grouped into three categories based on their CAE. These categories are most efficient, efficient and least efficient, having CAE score more than 2.94, between 2.94 and 1.62 and less than 1.62 respectively. Table-2 Figure-3 present the percentage distribution of extension personnel of different sub-systems under these categories.

Table-2. Sub-system-wise percentage distribution of EP to different efficiency groups.

Sub-systems	Most-efficient	Efficient	Least efficient
Extension specialists	14.3	28.6	57.1
Asst. Ext. Specialists	14.2	33.9	51.9
Zonal officers	14.2	42.9	42.9
Area officers	40.0	33.3	26.7
Agrl. Inspectors	50.5	32.3	17.2

There is very little difference in the percentages of efficient Extension Personnel of all sub-systems. This varies from 28.6 to 42.9

percentage. The percentages of most efficient Extension Personnel belonging to three sub-systems i.e. ES, AES and ZO are the same and as low as 14.2. In case of AO and AI the percentage of most efficient Extension Personnel is comparatively much higher. However, there exists a wide variation in percentages of least efficient Extension Personnel of all sub-systems. In general, sub-system which has higher percentage of most efficient extension personnel, has comparatively lower percentage of least efficient extension personnel. Slight variation to this generalization can be observed only in case of ZO which has equal percentage of efficient and least efficient extension personnel. Another generalization from the above finding is that majority of extension personnel, belonging to the university, fall into least efficient to efficient group, whereas majority of extension personnel, belonging to the Department of Agriculture, are in most or just efficient group.

Validity

The reliability and validity of the present measure of Communication-Act-Efficiency has not been attempted separately. However, since the communication efficiency is the degree of effectiveness, it promotes or facilitates relevant to the cost involved which is obviously observable; the only element that requires validation is the measure of effectiveness. The communication effectiveness which has been used as an enumerator in the formula of CAE has been measured in terms of information output. This effectiveness, however, has also been measured separately by two different methods, i.e. power evaluated and administrative evaluated effectiveness. When the scores of communicative effectiveness by all the three methods were correlated, they were found highly significant as presented in Table-3.

Table-3. Correlation between communication-act-efficiency and evaluated effectiveness.

Methods of evaluating effectiveness	Effectiveness and peer output indices	Communication act-efficiency	Peer evaluated effectiveness
Peer evaluation	236**	.019	
Administrative effectiveness	.305**	.039	.549**

**Significant at .01 level of probability

This suggests that the information output which has been taken as a measure of communication effectiveness, while conceptualising a measure for communication effective ness, while conceptualising a measure for communication-Act-Efficiency, is a valid one.

Both peer evaluated and administrative evaluated effectiveness have indicated non significant correlation with final score of Communication-Act-Efficiency. This finding has much to highlight the method of common evaluation itself. The non-significant correlation with CAE and significant correlation with information output are indicative of the fact that variables like cost, facility and job commitment are not considered by extension personnel, while apprising the communication performance. This means, extension personnel consider that the effectiveness lies in making more output, irrespective of cost and facility they have.

CONCLUSION

A method to measure the Communication-Act-Efficiency has been developed and used in the study. It has been partly validated. The output indices which were used as a measure of effectiveness show highly significant correlation coefficients separately with peer and administrative evaluated effectiveness. But the non-significant correlation coefficients of Communication-Act-Efficiency, with those of peer and administrative evaluated effectiveness, suggest the lack of appreciation of cost incurred and facilities availed by extension personnel for performing the act. However, this method of measuring Communication-Act-Efficiency helps in categorizing extension personnel into different levels of efficiency and comparing one category of extension personnel with other group.

In the course of study, this measurement was found short of the generality with which it could be used for comparing communication efficiency, irrespective of communication situations and cadres of extension personnel. This requires further refinement. Moreover, its reliability needs to be studied elaborately before adopting, in a changed situation.

REFERENCES

1. AKHOURI, M.M.P. (1973). Communication Behaviour of Extension Personnel –An Analysis of Haryana Agricultural Extension System. Unpublished Ph.D. thesis, IARI, New Delhi, India.

2. BOONE, EDGAR J. (1959). The Professional Status of Extension Specialists as compared with Research Resident Teaching Staff of selected Departments in Four-Land-Grant Institutions. Unpublished Ph.D. thesis. Madison; University of Wisconsin, U.S.A.

3. BROWN ENVIORY J. & DEEKENS, ALBERT (1958). Roles of Extension Subject Matter Specialist, Rural Sociology, 23, pp. 263-273.

4. CHAKRAVORTY, T.K. (1971). A Study of Factors Associated with Job Satisfaction of Block Extension Personnel in two IADP Districts of Northern India: Unpublished Ph.D. thesis, IARI, New Delhi, India.

5. DUBE, S.C. (1958). India's Changing Villages. Roultedge and Kegan Paul Ltd., 1958.

6. RAHUDKAR, W.B. The Relationship of Certain Personal Attributes to the success of Village Level Workers. The Indian Journal of Social Work, 1963, 23(4).

7. SALVI, P.V. & DUDHAN1, G.M. (1967). Role of Personal Characteristics in the Job Effectiveness of Village Level Workers, Indian Journal of Extension Education, 3(3) p. 127.

8. SENGUPTA, T. (1966). Characteristics of Effective Village Level Workers. Unpublished M.Sc. thesis, Division of Agricultural Extension, IARI, New Delhi, India.

9. SINGH, K.N. & PRASAD, N. (1972). A Study of Inter and Intra Departmental Communication. Indian Journal of Extension Education, 8 (1& 2), p. 1.

10. THAYER, L. Communication and Communicative Systems. Homewood, Illinois: Richard D. Irwin, Inc.

Chapter **4**

Communication Behaviour of Subject Matter Specialists in the Training and Visit Extension System

S.N. Pandey and R.N. Mishra

Systematic linkage amongst the research, extension and the client systems is the *sin-qua-non* organic part of the T&V system. The foremost requirement for the communication effectiveness is to take correct decision in selecting appropriate production technology well consistent with the agro-climatic and the socio-economic set-up of farmers of the area of system's operation *vis-a-vis* the technology being potent enough to show direct economic gain. Similarly, as the system envisages, the technology also needs to start with its simplest form and gradually move towards its most sophisticated form depending upon the corresponding increase in the technology-integration-potentiality of the farmers and improvement in their knowledge and skill about it. To meet this objective, the system requires adequate interaction with the various sources of information inputs, a careful information processing and taking precise judgment to select relevant production technology.

Thus, the role of subject matter specialists in the T&V System appears to be very important as they are the key personnel given responsibility to acquire process and screen the appropriate communication contents. Besides these, they have to discharge the other important functions like onward transmission of the selected farm messages to the field extension workers through periodically fixed programme of fortnightly trainings and providing them technical guidance in the field. They have to also plan and supervise the field trials to examine the economic profitability and adaptability of the recommendations and to ensure regular feedback from the fields through field visits and making interpersonal contacts with the farmers. Keeping these ideas in view, a research project was designed to analyse the communication behaviour of the subject matter specialists working under the T&V Extension system in terms of acquisition, evaluation, preservation, transformation and dissemination of farm information.

Methodology

Location and Respondents

The study was conducted in the Chambal Command Area Development Project of Kota in Rajasthan. The CAD Project of Kota is one of the three places where the T&V system was initially started on the experimental trials as the World Bank Project during June, 1974. The other two places were the Chambal CAD Project of Madhya Pradesh and the C.A.D. Project of Andhra Pradesh. There were two distinct categories of the S.M.Ss. working under the system *i.e.* (a) the Project level or Senior S.M.Ss. posted at the Project Headquarter, and (b) the District level or Junior S.M.Ss. located at the District levels. There were four senior specialists - one in each of kharif crops, Rabi crops, plant protection and farm management. The whole Project area was divided into three districts, namely, Bundi, Baran and Sultanpur. In each district, there were three Junior specialists in the subjects of crops, plant protection and water management. Thus including all, there were 13 specialists which constituted the respondents under the study.

Operationalisation of Concepts

Information Input Behaviour : It was operationalised as the activities performed by an individual Subject Matter Specialist for acquiring scientific farm information related to paddy from various sources. The amount of information input was measured by developing an information input index.

Information Processing Behaviour : When a scientific information is acquired from a source, it is not outright accepted for onward transmission to the extension workers or the farmers rather it is processed adequately to lest its adaptability in terms of existing agro-climatic conditions, crop production potentiality of the farmers, economic profitability and other relevant aspects. Empirically, it was defined as such activities performed by an individual S.M.S. for evaluation, preservation and transformation of acquired scientific information related to paddy before their onward transmission in the area. A standardised informationprocessing index was developed to measure the amount of processing made by an individual Subject Matter Specialist.

Information Output Behaviour : It referred to all such activities performed by an individual S.M.S. for disseminating scientific farm information on paddy among the extension workers and the farmers of the project area. The amount of information output was measured with the help of information output index developedunder the study.

Communication Behaviour : It was operationally defined as the composite of all the activities performed by the specialists for acquisition, processing and disseminatión of the Scientific farm information related to paddy. It was analysed at individual level in terms of information input, information processing and information output behaviour of the Subject Matter Specialists.

Findings and Discussion

The findings have been presented in three parts:

(A) Information Input Behaviour, (B) Information Processing Behaviour, and (C) Information Output Behaviour.

(A)Information Input Behaviour

Differential Use-extent of Sources and Channels : Table 1 reveals that the Senior S.M.Ss. utilized all the sources/channels under the study for their information inputs, whereas, the Junior S.M.Ss. not at all used the sources like affiliation to scientific societies, reprints, research seminars/workshops, discussion with researchers, making visit to research centers located outside the project area and attending the professional meetings held at the Project Headquarters. The most utilized sources/channels by the Senior S.M.Ss, were the professional meetings (95.1 %) followed by insights gained from the results of field trials (85.9%). The use-extent of research journals and other publications, problems received from the Junior S.M.Ss. visit to research centers located in the project area, discussion with researchers of the project research centre, problems received from the extension workers and farm broadcast were found to be moderate (per cent scores between 33 and 66). The remaining sources such as affiliation to scientific societies, reprints, campaign meetings, research seminars/ workshops, problems received directly from the farmers and newspapers were used to the least extent.

Table-1. Average use-extent* of different sources and channels for information input indices of subject matter specialists

Sources and channels	Senior specialists	Junior specialist
Affiliation to scientific societies	9.4	0.0
Research publication	50.4	28.3
Extension publication	48.9	32.0
Reprints	5.0	0.0
Campaign meetings	24.5	16.7
Inservice training	45.4	80.0
Research seminars/symposia/conferences/workshops	5.1	0.0
Consultation with scientists/specialists of GADP	47.0	30.0
Consultation with scientists/specialists of other than GADP	10.1	0.0
Visit to research and experimental centers within the GADP	47.5	30.5
Visit to research and experimental centres other than the GADP	8.1	0.0
Problems received for solution through –		
(a) Extension personnel	40.7	40.7
(b) Farmers	10.5	16.7
(c) Subject matter specialist	48.3	10.5
Broadcast/telecast	44.9	44.5
Newspapers	27.0	15.5
Professional meetings	95.1	0.0
Field trails	85.9	65.8

*the sum of sue-extent score in more than 100 due to multiple responses.

The senior S.M.Ss. were squarely responsible to acquire process and finally decide the communication contents for the whole project area in their respective areas of specialisation. They were also entrusted with the responsibility to impart fortnightly trainings for the field extension workers in all the districts with a view to maintain the homogenity of farm recommendations. In view of such functions, the senior specialists are utmost required to be in close touch with the researchers and also with relevant literature for being well informed about latest technical know-how. But the finding revealed that their exposure to such important sources was not up to the mark.

Their interaction was mainly confined to the researchers of the project research farms through bi-annual meetings of the Research and Extension Committee held usually before the onset of *Kharif* and *Rabi* seasons. It was also found that neither officially nor on their own initiative they interacted with the scientists of the research stations located outside the project-area and as such, the extension system as a whole, did not have any formal linkage with State Agricultural Universities or with National Research Institutions. The system also did not provide adequate library facilities for general use of the Project staff.

Nevertheless, participation in various types of professional meetings for discussion on matters related to problems and progress in agril. extension works, coordination, deciding training contents for the fortnightly training sessions, administration etc. is the most routine type work for the senior S.M.Ss. Similarly, conducting field trials for testing the economic profitability and adaptability of the new recommendations in the farming system of the area from the special features of the system which, in turn, serves an important source for gaining new insights and feedback for them. These may be the obvious reasons for the maximum use-extent of these two sources for receiving information inputs.

The informal discussion with S.M.Ss. revealed that they were not properly facilitated to visit research centers located outside the project area for increased interaction with the researchers and also for participating in seminars/workshops/symposia organized time to time either at the state or the national levels. Moreover, inadequate interaction with the local researchers indicates lack of proper initiative and interest on the part of specialists.

As regards district level subject matter specialists, they were found to be mainly dependent on fortnightly trainings imparted by the Senior specialists and the insights gained from field trials conducted by them in the area. As it was found, the T&V system at Kota did not provide separate arrangement for the training of the S.M.Ss. All the Junior specialists had to attend the fortnightly training sessions jointly alongwith the AEOs and VEWs. Secondly, to conduct and supervise field trials formed a part of their routine type work. These may be some of the obvious reasons for the maximum use-extent of those sources. However, these specialists not at all used the sources like affiliation to scientific societies, reprints, research seminars/workshops, discussion with researchers located outside the project area, visit to research centers located outside the project area and the professional meetings organised time to time at the district levels. The use-extent in case of receiving farm problems from the AEOs and VEWs and farm broadcast was found to be below moderate, whereas the remaining sources such as use of research and extension publications, campaign meetings, visit to the project research centre, problems received from farmers, discussion with colleagues and Senior specialists and newspapers were very poorly used. There are various reasons to support such findings, (a) The T&V system of Kota did not provide library facility or an information cell at the district level which might subscribe relevant literature for the use of the extension staff. They also lacked proper initiative and interest to subscribe for these publications. (b) Since such S.M.Ss. were more in touch with the field extension workers, they might have received more problems through them. They were also found neither trained in their subject areas nor systematic arrangement was made to facilitate their visits to research centers for increased interaction with the scientists. However, a few casual contacts with the scientists of the project research centre were mentioned by them which were made during the training sessions organised on such centre alongwith the AEOs and the VEWs.

Information Input Amount : The information input indices of the specialists ranged from 40.1 to 79.5 per cent, the system's mean input amount being 57.5 per cent. The t-test revealed significant difference between the average input-indices of the senior specialists (65.5%) and those of the Junior specialists (49.5%). This may be due to the obvious reason that the senior specialists were responsible for formulating the communication-contents for the project supposed to be exposed to various sources of information inputs.

(B) Information Processing Behaviour

It comprises of all the activities of a specialist undertaken after getting a farm problem. It has been presented in terms of exploring probable solutions, innovation evaluation, innovation preservation and innovation transformation.

Exploring Probable Solutions : It was found that almost all the Senior specialists conducted field trials and surveys, discussed with the Scientists of the project research centre, discussed with fellow specialists, discussed in project committee meetings, consulted literature and discussed with the farmers and the extension workers.

None of them could discuss with researchers located outside the State and only a few of them (1.85%) discussed with the specialists of the State Department of Agriculture and the State Agricultural University. As regards the Junior specialists, none of them discussed with the scientists of the project research centre, State Department of Agriculture, State Agricultural University or the research institutions located out side the state and also did not discuss in the meetings of project level committees. The average Junior specialists consulted literature for exploring a probable solution.

Innovation Evaluation : The system envisages that after a probable solution is arrived at, it should be subjected to rigorous evaluation. It was found that all the senior specialists evaluated a farm innovation through discussing it with the fellow specialists and also in the meetings of the project level committees. A fairly high percentage of them analysed the pros and cons of innovation in the light of past experiences (75.85%) and existing socio-economic and agro-climatic conditions (90.92%), considered the technical feasibility (65.23%), examined the validity of recommendations by conducting field trials (80.78%) discussed with the field extension workers (96%), and cross-checked against the past recommendations (98.02). None of them could discuss with specialists located outside the state and only a few discussed with the scientists of the State Agricultural University (5.85%) as well as the contact farmers (5.25%).

As regards the Junior specialists, out of the 12 methods of innovation-evaluation under the study, they did not discuss with the scientists of the project research centre, specialists of the State Department of Agriculture, State Agricultural University and the specialists located outside the State. But all of them found discussing with fellow specialists and also a high percentage of them discussed with extension workers (95.35%) and examined the validity of recommendations in view of the field trials (78.78%). Majority of them evaluated in the light of existing socio-economic and agro-climatic condition (67.85%) and the technical feasibility (67.85%). A few of them used the methods like cross-checking against past recommendation, analysing in the light of past experiences and discussing with the contact farmers.

Innovation preservation : The specialists did not depend much upon their memories and all of them preserved the farm messages by maintaining subjectwise files followed by noting in specified note-book by the Senior specialists (61.2%) and noting in a common notebook by the Junior specialists (68.89%). None of the specialists used reference cards for information preservation.

Innovation transformation : On ascertaining the usefulness of an innovation, it needs to be transformed in such forms so that the language of science may be easily comprehensible to the extension workers and the farmers. The T & V system envisaged originally that for creating effective learning situation, preparation and use of visual aids, distributionof leaflets/handouts containing impactpoints of the recommendations made in the training sessions, arranging method demonstrations etc. need to be more emphasized.

Table 2 reveals that all the senior specialists used the methods like preparing lecture notes for fortnightly trainings, developing leaflets/handouts and writing progress reports of the extension activities. Majority of them used the methods such as developing success stories (75.85%), circular letters (75.35%) and preparing package of practices in their areas of specialisation (53.25%). Transforming the innovations into the forms of charts, graphs and other visuals, news-stories and cyclostyling the full package of informations were practiced by a lower percentage of them.

Table-2. Percentage* of subject matter specialists using different methods of innovation transformation

Methods	Senior specialists	Junior specialists
Lecturer notes	100.00	60.35
Charts, graphs and other visuals	40.25	30.00
Leaflets/handouts	100.00	0.00
Cyclostyling whole recommendations	15.25	0.00
Success stories	75.85	44.44
Reports	100.00	78.44
Radio talk	0.00	0.00
Circular letters	75.35	24.44
News stories	20.255	10.55
Preparing simple package of practices	53.25	0.00

*The sum of percentage is more than 100 due to multiple responses.

Looking into the pattern of information transformation of the Junior specialists, writing reports (78.44%) and preparing lecture-notes (60.35%) were found to be the most used methods, preparing leaflets, getting the whole recommendations cyclostyled, writing radio scripts and also package of practices of crops were not at all practiced by them. The remaining methods such as transforming the farm innovations into visual aids, success or news stories and circular letters were used by a few of them. Poor use-extent of the innovation preservation techniques like visual aids, success and news stories suggests that for generating meaningful learning experience among the extension workers as well as the farmers, these techniques need to be widely used.

Information Processing Amount : The processing indices of the specialists ranged from 45.1 to 85.9 per cent, the system's mean processing amount being 61.2 per cent. The t-test revealed that the average processing amount of senior specialist (72.3%) was significantly higher than those of Junior specialists (50.1%). This is primarily due to the reasons that in the T&V system of Kota only the senior specialists were made responsible to acquire, process and finally select the relevant production messages for the entire project area.

(C) Information Output Behaviour

An innovation after its rigorous evaluation as well as its transformation into comprehensible forms gets ready for onward transmission to the extension workers and the farmers. Scientific information permeates to the farmers directly from the specialists and it is also funneled through the extension workers in the T & V system. As the system envisages, the subject matter specialists have to devote one-third time to field visits and making contacts with farmers besides their functions related to training and linkage with research. In the CAD project of Kota, where the district level S.M.Ss. were not associated with conducting fortnightly training sessions, they were to mainly devote most of their time in supervising the field trials, providing technical guidance to the AEOs and the VEWs, and making contact with the farmers.

Differential use-extent of methods I channels for innovation dissemination : Table 3 reveals that the most popular methods and media for innovation dissemination among the senior specialists were the fortnightly trainings (99.8%), professional meetings (90.6%), field trials (90.1%), field days' lectures (72.1%), releasing leaflets/handouts (68.9%) and drawing package of practices (67.5%). The use-extent of methods like field visits and contact with extension workers (43.9%), organising film-shows (48.5%), developing visual aids (49.2%) and group meetings with farmers and extension workers (31.8%) was found to be below average. Similarly, the methods such as field visits and contact with farmers, research seminars/workshops, training to the farmers, writing advisory letters to extension workers, telephone calls and releasing success or news stories were used to poor extent. However, the remaining five methods were not at all used by them.

As regards junior S.M.Ss., out of 21 methods/media the most used ones were: professional meetings (90.2%), field trials (89.8%) and making contact with extension workers (67.5%). Field visits and contacts with farmers were moderately used, whereas the use-extent of agri-films was below the average. The remaining used methods such as group meetings with extension workers and the farmers' trainings, field days' lectures, telephone calls, success or news stories and use of visual aids were used to the least extent. The other methods were not found to be in practice.

Conducting fortnightly trainings in the whole project area with a view to maintain homogenity of agri-recommendations, field visits, supervising field trials, writing package of practices on selected crops, preparing leaflets on the impact points of recommendations made in the fortnightly trainings, delivering field-days' lectures and participating in different professional meetings held at the project level, making linkage with researchers of various research centres/institutions etc. form the important functions of the Senior S.M.Ss. Thus, the maximum use-extent of many such methods is self-explanatory. Moreover, the poor use-extent of the methods such as conducting group-meetings with farmers and the extension workers, farm visits and making contacts with farmers, attending seminars/workshops/symposia, organising occasional trainings for farmers, releasing news or success stories, writing advisory letters to extension workers etc. suggest that the senior S.M.Ss. also need to be motivated to improve their performance in these important areas as the system envisaged.

The senior specialists visited the area only when they were informed about certain technical problems in their areas of specialization. The least use of writing advisory or circular letters to the extension workers and making telephone calls might be due to the reason that the extension workers or the farmers might not have sought guidance or advice either through writing letters or on telephone. Moreover, telephone facilities were also not available to the extension workers or the farmers.

Writing for magazines or bulletins, contributing scientific papers, radio scripts and organising agri-fairs, and exhibitions were not at

Table-3. Use-extent of different methods* and media by subject matter specialists

Methods	Senor specialists	Junior specialists
Fortnightly training sessions for extension personnel	99.8	0.0
Group meetings with farmers and extension personnel	31.8	32.7
Farm visit and contact with farmers	19.5	50.1
Contact with extension personnel	43.9	67.5
Seminars/symposia/conferences/worksh ops	25.1	0.0
Demonstration/minikit trail	90.1	89.8
Training to farmers	12.5	28.3
Field day's lectures	72.1	20.3
Drawing up package of practices	67.5	0.0
Professional meetings held at the project/district level	90.6	90.2
Advisory letters to farmers	0.0	0.0
Advisory letters to extension workers	25.3	0.0
Telephone calls	20.3	7.3
Leaflets/handouts	68.9	0.0
Magazines/bulletins	0.0	0.0
Sciencintifc papers/popular articles	0.0	0.0
Success stories/news stories	25.8	10.3
Film shows	48.5	40.1
Broadcasts/telecasts	0.0	0.0
Developing visuals	49.2	20.8
Farmers' fairs	0.0	0.0

*The sum of methods is more than 100 due to multiple responses.

all used by the specialists. This might be due to the reason that T & V system originally did not rely much on mass media for motivating the farmers.

The Junior specialists not at all used a few other methods such as conducting fortnightly trainings, attending seminars, workshops, drawing up package of practices and releasing leaflects or hand outs. Similar reasons as explained in case of the Senior specialists may also be applicable in the cases of these specialists. However, the Junior

S.M.Ss. were not at all associated with imparting fortnightly trainings and they mostly used the methods directly related to their most routine type jobs like supervising field trials, attending professional meetings and contacting the field extension workers. The use-extent of farm visits and making contact with farmers was higher among the Junior specialists as compared to senior specialists. This is due to the reason that they were supposed to devote most of their time in the field whereas the senior S.M.Ss. were to devote only a part of the time for such purposes.

Conclusion

Systematic linkage of the Subject Matter Specialists with various sources of farm information provides the T & V system its startup force and is a basic requirement for proper processing and transmission of relevant technology to the extension workers and to the farmers. The study revealed that the project level Subject Matter Specialists lacked formal as well as informal linkage with the information input-sources external to the project area. They poorly interacted with the local researchers of the project research centre, were less exposed to scientific literature and other mass media sources of farm information and kept themselves aloof of the seminars, symposia, workshops etc. organised either at the State or national levels. For their information inputs, the district level S.M.Ss. mainly depended on the production recommendations made by the senior S.M.Ss, in the fortnightly training sessions and the insights gained from the field trials undertaken by them in the project area.

Similarly, the information processing and output functions of the Subject Matter Specialists were also not up to the mark, the system originally envisaged in the CAD project of Kota. The output functions related to field visits, interpersonal interaction with the extension workers and the farmers, preparation and use of visual-aids, conducting method and result demonstrations, timely releasing the leaflets containing impact points of the training lessons and their distribution, integrating mass media with the system etc. specially need much improvement. The study implies that for efficient job performance, the Subject Matter Specialists need to be given adequate foundation training in their areas of specialisation, proper information support through creating information cell both at the project and district levels. They also need to be formally linked with the various research censers located within as well as outside the State. Mobility of the specialists is also one of the deficient job-areas which require proper organizational support as well as an increased morale on the part of the specialists. The study further implied that despite the organizational support, the personal initiative and interest of the specialists is also a vital factor to decide the communication behaviour in right perspective.

Chapter **5**

Information Output Pattern of Agricultural Researchers

Subhashchandra, S. Patil and M.B. Channegowda

Farm visits, office call by farmers, advisory letters to the farmers, office call by extension personnel and contacting salesman or dealer are the individualcontact methods; demonstrations, field day lectures, professional meetings and group meetings are the group contact methods; while publications and radio broadcast are the mass contact methods employed by the researchers of the Regional Research Station, Dharwad in Karnataka for information output. There is no significant difference in the frequency of use of methods for information output between the researchers of the mainstation and substations of the Regional Research Station, Dharwad in Karnataka.

With the ever increasing demand for food due to increase in population, the need for modernization of the agricultural sector becomes more important. Even when the technologies are available, unless the research information delivery system is effective, the technologies will remain on the shelves and in the files of research organizations, without making the expected impact. Gone are the days when the absence of technologies was complained of. Now the problems in most developing countries are those of adopting the available technologies to suit the situation and particularly the problem of effective delivery. The best way to start is by analysing the situation and diagnosing the problems so that appropriate solutions can be prescribed.

The University of Agricultural Sciences, Bangalore which was established on 22-8-1964 has five Regional Research Stations and 34 substations. A school of thought projects that the information output pattern of researchers at the main research stations is better than the researchers at the substations. This does not find objective evidence in the research stations of Karnataka including the Regional Research Station, Dhanvad of the University of AgriculturalSciences, Bangalore. Therefore, the present study has been undertaken with the broad objective of identifying the information output pattern of researchers at the main station and at the substations of the Regional Research Station, Dharwad in Karnataka State.

Methodology

The study was conducted during the year 1982-83 in the Regional Research Station, Dharwad which has a main station at Dharwad and eight substations spread over Dharwad and Belgaum districts of Karnataka State. All the researchers in the cadres of Professor, Associate Professor, Assistant Professor and Instructor excluding those in basic sciences and humanities, who had served for atleast six months as researchers were selected from the substations and eqûal number of eligible researchers were selected at random from the mainstation. The sample size was 70.

The information output pattern of researchers included (i) methods/ channels used by the researchers and (ii) frequency of use of methods/ channels by the researchers. The data were collected presonally by a pretested interview-schedule. CM square test was used to compare the information output pattern of researchers of the main station and substations of the Regional Research Station, Dharwad.

Findings and Discussion

The finding of the study and their discussion have been presented under the following headings: *

1. Methods used by the researchers.
2. Frequency of use of methods by the researchers.

Methods used by the Researchers

Farm visits, office call by farmers, advisory letters to farmers, office call by extension personnel and contacting salesman/dealer were the individual contact methods employed by the researchers (Table 1). Demonstration was the most used method followed by field day lecture, professional meeting, group meeting, farmers' fair, visit to farmers organizations and kharif and rabi campaign meetings among the group contact methods. Publications and radio broadcast were the mass contact methods employed by the researchers. There was similarity in the use of methods for information output by the researchers of the main station and the substations of the Regional Research Station, Dharwad.

Table-1. Different methods used by researchers in the process of information output

S. No.	Methods	Researchers					
		Main station (N=35)		Substations (N=35)		Total (N=70)	
		No.	%	No.	%	No.	%
Individual contact methods							
1.	Farm visits	30	85.7	32	91.4	62	88.6
2.	Office call by farmers	26	74.3	27	77.1	53	75.7
3.	Advisory letters to farmers	24	68.6	26	74.3	50	71.4
4.	Contacting salesman/dealer	23	65.7	21	60.0	44	62.9
5.	Contacting salesman/dealer	20	57.1	22	62.9	42	60.0
6.	Telephone call by farmers	21	60.0	17	48.6	38	54.3
7.	Advisory letters to extension personnel	18	51.4	15	42.9	33	47.1
8.	Telephone call by extension personel	20	57.1	13	37.1	33	47.1
Group contact methods							
1.	Demonstrations	33	94.3	31	88.6	64	91.4
2.	Field day lecturers	30	85.7	31	88.6	61	87.1
3.	Professional meetings	30	85.7	30	85.7	60	85.7
4.	Group meetings	29	82.9	26	74.3	55	78.6
5.	Farmers fair	21	60.0	25	71.4	46	65.7
6.	Visit to farmers organizations	22	62.9	20	57.1	42	60.0
7.	Kharif and rabi campaign meetings	20	57.1	22	62.9	42	60.0
8.	In service training meetings	20	57.1	18	51.4	38	54.3
	Mass contact methods						
1.	Publications	32	91.4	60	85.7	62	88.6
2.	Radio broadcast	19	54.3	21	60.0	40	57.1

The results of the present study in respect of methods employed by the researchers for information output are in agreement with the findings of Rao (1972), Ambastha and Singh (1977), Chamala (1979), Dudhani (1980), Lakoh and Akinbode (1981), Thitai (1981), Akinbode (1982), Channegowda (1982) and Rockett and Smith (1982). As for the similarity in the use of methods for information output by the researchers of the main campus and substations of the Research Institute, the findings of the present study conform with the findings of Sanoria (1978) in respect of group contact methods and do not conform with individual contact methods and mass contact methods.

The possible reasons for large percentage of researchers of the Regional Research Station, Dharwad in using the above mentioned methods may be due to the effectiveness of methods accessibility, organizational norms and conditions and personal interest of the researchers to use these methods.

Frequency of use of Methods by the Researchers

There was no association between the researchers of the mainstation and substations of the Regional Research Station, Dharwad with their frequency of use of methods as revealed by the non-significant results of the chi square test (Table 2). The variation observed in the data between these two categories of researchers was due to chance factors and not due to the real difference.

Table-2. Frequency of use of methods employed by researchers for information output

S. No.	Frequency use	Researchers					
		Main station		Substations		Total	
		No.	%	No.	%	No.	%
1.	More frequently	16	45.7	10	28.6	26	37.1
2.	Frequently	5	14.3	10	28.6	15	21.4
3.	Less frequently	14	40.0	15	42.8	29	41.5
	Total	**35**	**100.0**	**35**	**100.0**	**70**	**100.0**

$\chi^2 = 3.08$, Df = 2, Not significant

The possible reasons for the similarity in the frequency of use of methods to disseminate research information by the researchers of the main station and the substations may be due to the availability of the methods equal opportunities to use the methods. Similar need of the methods, specialization of researchers in the substations and similarity in the ability of researchers of the main station and substations in using the methods of information dissemination.

Implications

The communication methods such as (i) demonstrations, (ii) farm visits, (iii) field day lectures, (iv) professional meetings, (v) group meetings, (vi) office call by farmers, (vii) advisory letters to farmers,

(viii) office call by extension personnel and (ix) farmers' fair were extensively used by the researchers for information output. This implies that the organization concerned with research will have to promote the effective methods as mentioned above for use by researchers for information output.

Similarity in the information output between the researchers of the main station and the researchers of the substations imply that adequate opportunities in the use of communication methods given to the researchers of the substations will enable them to measure upto the level of the researchers of the mainstation. Besides, encouraging specialisation is another important element which enables researchers of substations to grow and develop as those of the researchers of the mainstation.

REFERENCES

1. AKINBODE, A., "A critical analysis of the management of agricultural extension in Nigeria," Agric. Adm., 10(1) : 54-55, 1982.
2. AMBASTHA, C.K. and SINGH, K.N., "Communication pattern of farm scientists : a system analysis," Indian J. Ext. Educ., 13 (1&2): 22-30, 1977.
3. CHAMALA, S., "Agricultural extension and China's rural development," *Rur. Dev. Dig.*, 2(2): 39-52, 1979.
4. CHANNEGOWDA, M.B., "Interaction between researchers and farmers in Bangalore district of Karnataka," Unpub. report, Department of Agricultural Extension, Uni. Agril. Sci., Bangalore, 1982.
5. DUDHANI, C.M., "A study on the linking role of extension supervisors, subject matter specialists and research workers," Ph.D. Thesis (Unpub.), Uni. Agril. Sci., Bangalore, 1980.
6. LAKOH, A.K. and AKINBODE, I.A., "Agricultural research delivery system in Sierra Leone," Agric. Adm., 8(4): 289-305, 1981.
7. RAO, R.C., "Communication linkages in transfer of agricultural technology", Econ. and Political Wkly., 7(53) : 157-70, 1972.
8. ROCKETT, B.W. and Smith, J.T., "Interaction between state funded agricultural research and development establishments and agricultural education : case studies in England and Wales," Agric. Adm., 10(4) : 265-84, 1982.
9. SANORIA, Y.C., "Communication efficiency of agricultural scientists," Indian J. Ext, Educ., 14(3 & 4) : 81-83, 1978. THITAI, G.N., "Transfer of research results Production and dissemination of research results in Kenya," National Research Centre; 91-93, 1981.

Chapter 6

Communication in Training and Visits Extension System

P.N. Mathur and S.N. Pandey

The present paper makes a concise attempt to throw light on a few communication aspects of the project (training and visits system) like communication behaviour of the personnel, information gap and feed back mechanism among research, extension and client system. The study was carried out in Kota (Rajasthan) during 1978-79 for which samples were drawn from subject matter specialists, field extension workers and farmers of the project area growing paddy. The study indicated that the farmers were mainly using personal locality sources for farm information, the extension workers were found to maintain no interaction with research institutions within and outside the project area and the interaction of the project level SMS was also mainly confined to the project research centre. Significant information gap in all the important messages related to paddy were observed at different levels of extension and client system.

INTRODUCTION

Daniel Benor, the initiator of the Training and Visit Extension System, recently expressed that all is not well with this system and that it needs to be revitalized. According to him, new agricultural technology is all that is required to inject a new life in it. Problems of training, farm visit and contact with farmers, communication and personnel management, according to him, deserved lower priorities. This system was initially started in Chambal Command Area Development Projects of Kota in Rajasthan, and Madhya Pradesh, and command Area Development Project of Andhra Pradesh in June, 1974 with World Bank's assistance. Later on, keeping the recommendations of the National Commission on Agriculture in view, the Government of India took a vital decision to introduce this system in all the states of the country under a centrally sponsored scheme entitled "Reorganisation of Agriculture Extension Administration in the States" which provided 75 per cent financial aid to the implementing states. Twelve Indian states have already adopted the system and the others are keenly watching its progress. Since its introduction in India, this system has faced many challenges.

A study was carried out in the Chambal **C.A.D.** Project of Kota in the State of Rajasthan during 1978-79. The research design selected for this study was field survey in actual settings. The Subject Matter Specialists (SMSs) and the Field Extension Workers working in the T&V Extension System of Chambal C.A.D. Project of Kota and the farmers of the Project Area growing paddy, constituted the total population of the study.

Selection of Respondents : The total sample constituted five categories of respondents. All the 13 **S.M.Ss.** - four Senior S.M.Ss. and 9 Junior S.M.Ss. located at the Project and District levels, respectively, constituted the respondents in the first category. The whole Project Area was divided into three extension districts for the purpose of extension work. One Extension District, namely, Bundi was selected randomly for selecting the **A.E.Os., V.E.Ws.,** contact and non-contact farmers that form the respondents in second, third, fourth and fifth categories, respectively. Ten **A.E.Os.** who were associated with disseminating farm information on paddy were selected out of the 12 total **A.E.Os.** in the Bundi District. In the third category of the respondents, 50 **V.E.Ws.** were selected at the rate of five in each of 10 selected circles of **A.E.Os.** on the random basis.

A list of major paddy growing villages was prepared for each of the 10 **A.E.Os.** - circles and they were arranged in descending order in terms of area devoted to paddy cultivation. Then two villages from the top were selected in each of the **A.E.Os.** circles. Thus, in all, 20 villages were selected. Out of these villages, 100 contact and 120 non-contact farmers were randomly selected at the rate of five and six, per village, respectively. But due to unavoidable reasons, responses could not be obtained from five respondents in each of these two categories of farmers. Thus, 95 contact and 115 non-contact farmers constituted the fourth and fifth categories of respondents. Besides these, a complete village, namely, Matunda was selected purposively among such selected villages for the socio-metric study.

Procedure of Data Collection : Data were collected by interviewing the respondents personally with the help of structured schedules. The structured schedules were composed of various indices, tests and scales to measure the variables like information input, processing and output, communication behaviour of the S.M.Ss., extension workers and the farmers, information gap, attitude, feedback and inter-personal communication at the village level. All these indices, scales and tests were tested for their reliability and validity. The communication pattern analysis was done mainly at the individual level.

FINDINGS AND DISCUSSION

Communication Behaviour of Personnel

Communication in the T&V system is vital for its success. However, recent research findings indicated that the farmers and the extension workers in Rajasthan Chambal Command Area have been mainly using neighbours, Friends, family members and village extension workers as sources of farm information. Some of the least used sources had been the mass-media like radio, agri-films, exhibitions, leaflets, bulletins, magazines, newspapers etc., and other important extension methods like campaigns, visits to research centers, field days, etc. The A.E.Os., the S.M.Ss. and the scientists had been seldom used by the farmers.

The above findings have indicated that many extension methods-mass-media were playing limited role in T and V system. Benor says, "I am delighted to involve available media and media specialists in the efforts, but, I do not think that even TV can train farmers in field practices. Farmers need people to demonstrate and to check their progress and to help them take into account the special characteristics of their own plots and land. TV cannot be specific enough. I appreciate the media, but I regard them as helpers."

Emphasizing the importance of extension strategies for developing countries, Benor says :

"Highly specific and centrally developed message sent out in lieu of extension agents do not work. The same track works in the most highly agriculturally developed countries, where people need good ideas that reach them through the media and have the means to try these ideas. For countries between the extremes, other strategies have to be used."

It is evident that mass-media and several extension methods have not been put to use in the extension system. Even the extension workers have been mainly using fortnightly trainings and the hand-outs released during these trainings as the main source of information for themselves. Thus, the information input of the extension agencies has been limited to the training programmes only taking away all the initiative of the extension workers in reading agri-info-materials and accepting everything unreserved. Studies have shown that information input, processing and output ofthe extension workers and the contact farmers have been poorly resulting into information gaps at various levels. The extension agents and the contact farmers hardly resort to the use of established communication techniques of tapping as many sources of information as possible and using proper methods of recording the information for future use. This results into poor information input and a significant loss of message during its flow to the farmers.

As a sequel to poor information input and processing, the output of the extension workers has also been reported to be much below expectations. Besides its effectiveness, which ultimately depends upon the techniques of inter-personal communication, has been found to be low. This is perhaps one of the weakest links in this extension system.

The following table shows that the extension workers use only limited methods for information dissemination.

Table-1. Degree of use of various methods for information dissemination by extension workers (Kota).

S. No.	Methods used	Extent of use AEO	V.E.W.	Overall
1.	Demonstration/Minikit trail	High	High	High
2.	Visit and contact	Medium	Medium	Medium
3.	Field days	High	Medium	Medium
4.	Group meetings	Low	Low	Low
5.	Farmer's cells	Low	Low	Low
6.	Films	Low	Nil	Nil
7.	Other mass-media	Nil	Nil	Nil

The Project level Subject Matter Specialists were squarely responsible for acquiring latest scientific innovations, processing them and finally deciding the communication content for the whole project area. They also organized and imparted fortnightly trainings for the extension workers of the whole project area with a view to maintain homogeneity of scientific recommendations. As envisaged in the system, they were to devote their one third time in establishing close and regular interaction with the researchers of the research institutions located within and outside the Project area. They are, thus, required to be well-informed about the latest scientific innovations through research journals, books and other relevant publications. The study related to information input behaviour of such specialists revealed that the project level S.M.Ss interacted with the researchers of the Project Research Centre mainly in the Research and Extension Committee meetings which are held twice in a year before the onset of the Kharif and Rabi seasons; otherwise, on their own initiatives and interests, they had least contact with them. Usually, they contacted them when they sought technical help for finalising the communication-contentof the trainings and solving a specific problem in their area of specialisation which they could not solve. The use of research and extension publications was made to a moderate extent. They seldom participated in seminars/symposia/workshops etc. and consulted the scientists of the research centers located outside the C.A.D. Project.

The district level S.M.Ss were not at all affiliated to any scientific societies related to their subject areas and did not have any interaction with the research centers outside the **C.A.D.** Project. Their interaction with **C.A.D.P.** research station and use of research and extension publications was found to be very poor. They were mostly dependent on fortnightly trainings imparted by Project level S.M.Ss which they attended jointly with A.E.Os, and V.E.Ws. of the respective districts for acquiring scientific information. The ideas gained through self-observation of the adaptive researches conducted in the project areas served next important means for their information input.

The study further revealed that S.M.Ss were neither formally, nor on their own initiative established link with State level Agril Research Stations located outside the Project, and the extension system, as a whole did not have any formal linkage with the State Agril University or with any central level research institutions. It was further observed that the system did not provide adequate library facilities for general use of the Project staff and the extension service could not subscribe to relevant research publication for them. The S.M.Ss. stressed to develop information centers at the Project and District levels and also a mini-information cell at Panchayat Samiti levels.

Table-2. Per cent use-extent of different sources/channels for the information input of subject matter specialists.

Sources/Channels	Project level SMSs	District level SMSs
Affiliation to professional societies	9.0	Nil
Research publications	50.4	28.3
Extension publications	48.9	32.0
Reprints	5.0	Nil
Campaign meetings	24.5	16.7
Fortnightly trainings	45.4	80.0
Research Seminars/Workshops/Symposia etc.	5.1	Nil
Consultation with Scientists of the –		
(a) C.A.D. Project	47.0	30.0
(b) Other than C.A.D. Project	10.00	Nil
Visit to research centers –		
(a) Located within C.A.D. Project	47.5	30.5
(b) Located outside C.A.D. Project	10.0	Nil
Radio/T.V.	44.9	44.5
Newspapers	27.0	15.5
Professional meetings	95.5	Nil
Self-observation with adaptive research	85.9	65.8

Information Gap

Taking the issue of lack of techniques of communication forward, it has been found out that the information gap had been widening as the technological information flows from the A.E.Os. to the farmers. This situation has mainly arisen because of lack of communication skills on the part of the extension workers.

A further look at the information gap in paddy cultivation practices showed that plant protection measures, weed control, nursery preparation, etc. were some of the items in which more information gaps have been noticed. A look at Table 3 proves this fact:

Feedback between research, extension and client systems

A systematic feedback between research, extension and client system was also considered to be vital in the T & V system. But the findings revealed that farmers made feedback to only extension system which was almost confined to the Village Extension Workers.

Table-3. Hem-wise information gap in paddy cultivation technology in T & V system (Kota).

Level		Item and % gap	
AEO level	1.	Weed control	36
	2.	Nursery	34
	3.	Seed rate and seed treatment	32
VEW level	1.	Plant protection	32
	2.	Seed rate and seed treatment	31
	3.	Nursery	30
	4.	Weed control	26
Contact farmer level	1.	Weed control	94
	2.	Plant protection	83
	3.	Seed rate and seed treatment	68
	4.	Nursery	52
	5.	Transplanting	22
	6.	Fertilizer use	22
Non-contact farmer level	1.	Weed control	97
	2.	Plant protection	89
	3.	Seed rate and treatment	76
	4.	Nursery	26
	5.	Transplanting	26
	6.	Fertilizer use	17
Mean total information gap	1.	Non-contact farmers	54
	2.	Contact farmers	49
	3.	VEWs	24
	4.	AEOs	19

They were also not found to make purposive or deliberate feedback. Such feedback was sporadically made when asked by the field level extension functionaries like **V.E.Ws. and A.E.Os.** during their field visits. Similarly, in the extension system, only the Project level Subject Matter Specialists who were solely responsible to process, decide the communication-contents and conduct periodic trainings for the Project area, could only made feedback to the research system and it was mainly confined to the researchers of the research station of the project area. Most of the feedback was made during the discussion in the biennial meetings of Research and Extension Committee of which the researchers of the **C.A.D.P.,** Research Station and the Project level S.M.Ss. were members. Besides this, such S.M.Ss. contacted the researchers only when they felt special problems in their areas of specialisation which they were not able to solve. The S.M.Ss. located at the district levels did not make feedback to the researchers like **A.E.Os.** and **V.E.Ws.** and the farmers. They usually made feedback to the high-ups like Project level S.M.Ss. and the **D.E.Os.** when they were asked to furnish such information about the impact of advocated technology in the farmers' fields. They mostly conveyed such information in the fortnightly training sessions. The **A.E.Os.** and **V.E.Ws.** also, on their own initiative, did not make purposive feed-back within the extension system.

Attitude

The mean attitude scores of contact (2.58) and non-contact farmers (2.43) were found to be less than the neutral score (3.00) indicating that farmers in general had unfavourable attitude towards the T&V system. The average attitude score of **V.E.Ws.** was found higher than the neutral score by a very marginal difference (0.06) and as such the score was leaning towards neutral attitude. Only the mean attitude score of **A.E.Os.** (3.28) showed favourable attitude towards the system. The values of 't' and the underlined bars presented in Table-4 indicate that **A.E.Os.** and V.E.Ws. had higher attitude intensity than that of both categories of farmers. However, there were no significant differences between the **A.E.Os.** and V.E.Ws. and between the contact and non-contact farmers.

Table-4. Average attitude scores of non-contact farmers, contact farmers, V.E.Ws. and A.E.Os. and values of 't' between the means.

Mean attitude scores				Values of 't'					
Non-contact farmers	Contact farmers	V.E.Ws	E.	Non-contact vs contact	Non-contact vs V.E.W.	Non-contact vs A.E.O.	Contact vs V.E.W.	Contact vs. A.E.O.	V.E.W. vs A.E.O.
2.43	2.58	3.06	-3.29	-1.76	-6.43**	-4.21**	4.72**	3.44	1.07 NS

Selection of contact farmers

As regards the selection of contact farmers, the system originally envisaged that the contact farmers should be key-communicators and should be selected by seeking collective advice of the community. The findings revealed that such criteria were not fully adopted by the extension workers while selecting them. Table 5 indicates that all the contact farmers selected in the system were not key-communicators or even communicators, but a considerable percentage of them were non-communicators among their groups to which they belonged. In fact, there were also key-communicators and communicators among the other farmers of the groups which could not be detected.

Table-5. Position of key-communicators, communicators and non-communicators among the contact and non-contact farmers in percentage.

Levels of communication	Contact farmers (n=37)	Non-contact farmers (n=98)	Overal l
Key-communicators	24.32	8.16	12.59
Communicators	43.24	46.94	45.93
Non-communicators	32.44	44.90	41.48

Similarly, more than 92 per cent of the farmers including contact farmers had no knowledge about the selected contact farmers. Hence, it was found necessary that contact farmers need to be selected by taking joint decision of the farmers of the groups based on socio-metric method and they should represent all the sections of the farming community. They should however, be replaceable if the majority of the group so desired. It is also necessary to ensure that the contact farmers are made aware of their selection, role and responsibilities and that they are also motivated enough to render technical guidance to fellow farmers.

SUMMARY AND CONCLUSION

The study was conducted in Chambal C.A.D. Project of Kota in Rajasthan during 1978-79. The research design adopted was field survey in actual settings. The findings revealed that even after the operation of the T & V system in the Project area since 1974, the farmers were mainly using the personal localite sources like neighbours, friends, family members and village extension workers among the personal cosmopolite sources for seeking farm-information. Some of the least used sources were the mass-media like radio, agri-films, exhibitions, leaflets, bulletins, magazines, newspapers and other important extension methods like campaigns, visits to research centers, field-days etc. The **A.E.Os.**, S.M.Ss., scientists and other extension personnel had been seldom used by them.

Except Project level S.M.Ss., the District level S.M.Ss. and the extension workers had no interaction with research institutions within and outside the project area. The interaction of Project level S.M.Ss. was also mainly confined to the project research centre. There was no kind of formal linkage with state Agricultural University. The use-extent of research journals and other publications was found to be poor among the S.M.Ss.

Significant information-gaps in all the important messages related to paddy were observed at different levels of the extension and client systems. At the farmers' level, a wide gap was found in the messages related to weed control (more than 94%), plant protection (more than 83%) and seed rate and its treatment (more than 68%).

In the whole extension system, only the Project level S.M.Ss. made feedback to the research system which was usually confined to the researchers of the research centre located in the Project area. Farmers made feedback to only V.E.Ws. in the whole extension system. Such feedback was not deliberate or purposive.

In general, the farmers had unfavourable attitude towards the system, whereas, the V.E.Ws. had neutral attitude. The overall attitude intensity among A.E.Os. was found to be simply favourable. The selection criteria as envisaged in the system were not properly adopted for selecting the contact farmers.

Reference

BENOR, D. & HARRISON, J.Q. (1977). *Agricultural Extension : The Training and Visit System*, World Bank, Washington, D.C.

Chapter **7**

Communication Patterns of Agricultural Scientists- An Intra-System Analysis

Y. C. Sanoria and D.K. Sharma

An attempt has been made to analyse the communication patterns of agricultural scientists in an Agricultural University of India. Peer communication appears to play considerable role in influencing the researcher's information input and processing behaviour. In a multicampus situation, communication generally centralises around main campus. The senior scientists located at main campus play the role of opinion leaders in their respective fields of specialization.

The communication patterns of agricultural scientists can be analysed at three levels : (i) Individual communication patterns, (ii) Intra-system communication patterns and (iii) Inter-system communication patterns. At the individual level, in terms of systems parameters, the information input, processing and output have been examined. Communication patterns at the intra-system level can be separated into peer communication, opinion leadership and communication network centrality. At the inter-system level, the communication patterns may be isolated into communication transactions with extension workers and farmers. In this study, all these three levels of communication patterns have been examined, but emphasis has been, however, laid on intra-system level. At this level, it is of greater significance to understand how the agricultural scientists communicate with their peers of the same and different cadres? How communication networks are formed in an organisation? Who influences the opinions of the members and occupies the opinion leadership position in an organisational structure. The study attempts to answer such questions through critical examination and investigation of related issues.

Material and Methods

The study was conducted at Madhya Pradesh Agricultural University. This university has six agricultural colleges and seventeen research stations, located in various agro-climatic regions of the state, which carryout research, education and extension functions in an integrated manner. The university was conceptualized as "Research System" since it has both the interacting components and boundary with desired property as outlined by Berrien (1968). On the basis of organizational set up of agricultural research in the university, the system was isolated into two sub-systems, (i) the researchers working at the main headquarters, and (ii) the research stations and different campuses. As researchers were the main focus of the study, all the researchers working exclusively on the research establishment of the Directorate of Research represented the universe. For selecting the sample from the two sub-systems, a random sampling technique was used and the first 44 researchers from various departments were randomly selected. From the second sub-system, 40 researchers were selected at random. Thus, a total of 84 researchers constituted the sample for investigation. Data were collected through personal interview with the help of a structured schedule designed for the study. These were analyzed into standard scores and means and amount of the indices were worked out to determine the various parameters of communication patterns.

Operationalisation of Concepts

Communication Pattern : It was denned as the communication behaviour of an individual, dyad, or a social system that is systematic and exhibits some form of regularity. The systemic communication behaviour of an individual includes input, processing and output at the personal level; opinion leadership communication network and peer communication at the intra-personal level and communication with other departments and systems at the inter-system level.

Information input (X1) : It refers to the degree to which an individual uses various sources for getting research ideas and information in performing his role. This was measured by an index developed for the study.

Information processing (X2) : It refers to the degree to which an individual uses various processing methods like synthesis, evaluation, storage and transformation of research idea to perform his role. This was determined by an index prepared.

***Information Output* (X3)** : It was defined as the degree to which an individual disseminates research information with the help of various communication methods and media to perform his role. The amount of output was measured with the help of information output index.

***Peer Communication* (X4)** : Peer communication is the degree with which the individual communicates with his peers of the same or different cadres regarding the technical and professional matters. This was assessed by an index as suggested by Moreno (1953).

***Opinion Leadership* (X5)** : It is the degree, to which an individual is able to influence informally, other individual's attitudes or overt behaviour, in a desired way with relative frequency. This was measured with the help of Socio-metric Technique as suggested by Moreno (1953).

Communication network centrality (X6) : It is the degree, to which an individual occupies the central position, in the frequency of communication, with the seniors and juniors in the organisation related to the technical and professional matters. This was measured by Socio-metric technique of Moreno (1953).

***Researcher and other department colleagues communication* (X7)** : The variable refers to the frequency with which an individual communicates his own research results with other department colleagues. This was determined by an index prepared for the study.

***Researcher and extension personnel communication* (X 8)** : It refers to the degree with which an individual communicates scientific information to the member of extension system with relative frequency. This was also assessed by an index developed for the study.

***Researcher-Extension personnel contact span* (X9)** : It refers to the degree of contact an individual made to the extension workers for communicating scientific research information. This variable was again determined with an index formulated for the investigation.

***Researcher-Farmers Communication* (X10)** : It was defined as the degree to which an individual communicates farm information to the farmers by use of different communication methods and media. This too was determined by an index prepared for the study.

***Researcher-Farmers Contact Span* (X11)** : It refers to the degree of contact an individual had with the members of client system for communicating scientific information. This was assessed by a designed index.

Results and Discussion

The results of the study are presented in four parts: (i) Peer communication, (ii) Communication network centrality, (iii) Opinion leadership, and (iv) Researcher-other department communication.

(i) *Peer Communication*

The overall mean peer communication index of research system was 1.98. The sub-system wise mean amount of peer communication revealed that researchers of main campus had higher mean score (2.26) than those of research stations' researchers (1.66). The test, however, indicated (t=1.80) no significant differences in the peer communication amount of the; researchers of both the sub-systems. This may be due to the same amount of qualifications, interest in work and professional competence, due to which the researchers were found to be frequently communicating with one another about research work of their own field. These findings support the results of Pelze and Andrews (1966) who reported that scientists in research and development organisations and laboratories communicated with their peers equally and : their frequency was daily. The number of colleagues with whom they communicated varied from four to ten per persons.

Correlation with other communication indices : The correlation coefficients between peer communication index and other indices (Table 1) unfolds significant positive relationship with information input (r =.193), processing (r = .227), communication network centrality (r = .278), opinion leadership (r = .253), researcher-extension personnel communication (r = .264) and researcher-farmer communication (r = .245).

The peer communication index, however, did not show any significant correlation with output, researcher-extension personnel contact span and contact span with farmers and communication with other department. The positive and significant correlation of peer communication with the information input of researcher and processing is indicative of the fact that those researchers who communicate more with their peers, acquire more ideas for further research and when they have more ideas naturally they process more information and at the same time also discuss with the peers about their methods of processing. Further, some of the peers might be in the central position of communication network and also occupying the opinion leadership position and other researchers might be seeking their opinion and advice about the research work in agriculture. These findings are in line with the work of Vickery (1963) and Hirch *et al.* (1958).

Table-1. Correlation of Peer Communication Index with other Communication Indices of Researchers.

X4	X1	X2	X3	X5	X6	X7	X8	X9	X10	X11
Peer communication	.193*	.227*	.146	.253*	.278*	.004	.264	.015	.245*	.113

*Significant at .05 level of probability

Table-2. Correlation of Communication Network Centrality with other Communication Pattern Indices of Researchers.

X5	X1	X2	X3	X4	X6	X7	X8	X9	X10	X11
Communication network centrality	.296*	.283*	223*	276	.879*	.005*	284*	406*	254*	.165

*Significant at .05 level of probability

(ii) Communication Network Centrality (CNC)

The overall average communication network centrality index of research system came to be 0.78. The sub-system wise average communication network centrality indices were 1.23 and 0.76 for researcher of main campus and research station, respectively. The t-value of significance (t = 3.10) of difference revealed that both the sub-system researchers significantly differed in their amount of CNC. This shows that researchers of main campus of the Agricultural University hold central position in communication network to a greater degree than those of research stations. The obvious reason is that the researchers at the headquarters are senior persons, mostly possessing doctoral degrees and more experience.

Correlation with other Communication Pattern Indices – Table 2 shows correlation coefficients between communication network centrality and other communication indices. Examination of the figures reveal that CNC established significant positive relationship with information input (r = .296), processing (r = .283), output (r = .223), peer communication (r = .276), opinion leadership (r = .879), researcher extension personnel communication (r = .274), researcher-extension personnel contact span (r = .408) and researcher-farmers communication (r = .254).

The CNC had significant relationship with input, output and processing thereby emphasizing the crucial role played by CNC in determining the communication patterns. Similarly, the strong correlation with peer communication and opinion leadership suggests that researchers who communicate frequently with their peers might be in the central position and occupy the leadership and others might be seeking information and advice from such central figures. It is also revealed that CNC has relationship with researcher-extension personnel communication. Thus, it is evident that those researchers who occupy the central position also communicate with extension personnel and farmers frequently. Glanzer (1961), Singh (1972) and Jain (1970) reported similar findings.

(iii) *Opinion Leadership*

The overall mean opinion leadership index of research system was 10.65. The sub-system-wise average opinion leadership index was 12 in case of researchers of main campus and 8.67 for research station researchers. The t-value (t = 2.29) was significant at .05 level of probability, indicating that these researchers of the sub-systems differed significantly in their opinion leadership amount. The obvious reason for this difference may be that researchers at the main campus are senior persons holding higher positions in various departments and these persons influence and guide the research work, not only on the main campus but they also supervise the work of research stations.

Correlation with other communication pattern indices : Table 3 shows that, opinion leadership index had significant and positive correlation with input (r = .461), processing (r = .233), output (r = .303), peer communication (r = .398), communication network centrality (r = .792), researcher-extension personnel communication (r = -429), researcher-extension personnel contact span (r = .335), researchers farmers communication (r = .229), and communication with other departments (r = .255).

The positive and significant relationship of opinion leadership with input processing and output reveals that those researchers who are the opinion leaders acquire more information, process it and disseminate it either to extension workers or farmers evidently this has shown significant relationship with the communication and contact with extension personnel and farmers.

(iv) *Researchers' communication with other department scientists*

. The overall mean researcher other department communication index of research system was 6.80. The sub-system wise average were

8.08 and 4.76 for researchers of main campus and research stations, respectively. The l-test (t=3.29) indicated significant difference in the amount of index of the respondents of these two sub-systems. These findings support the work of Hirch (1958), Aronon (1968), Pelze and Andrews (1966), Lingwood (1969) and Lionberger and Chang (1970) who concluded that standard methods of communicating research results of scientists were: attendance in meetings; professional seminars and conferences, publications; personal contact and discussions.

Correlation between researchers-other department communication

Table 4 presents that researcher-other department communication index had positive and significant correlation with input (r=.315), processing (r = .221), output (r = .308), opinion leadership (r = .255), researchers-extension personnel communication (r = .409), researchers-farmers communication (r = .257) and researchers contact span with farmers (r = .387). The peer communication, communication network centrality and researchers -contact span were not found to be correlated with the researcher-other department communication index. Similar results were reported by Pelze and Andrews (1966).

Table-3. Correlation of Opinion Leadership Index with other Communication Pattern Indices of Researchers.

X6	X1	X2	X3	X4	X5	X7	X8	X9	X10	X11
Opinion leader-ship	461*	233*	303*	.398*	.792*	.255*	.429*	.335*	.299*	.199*

*Significant at .05 level of probability

Table-4. Correlation of Researchers-other Department Communication index with other Communication Pattern Indices of Researchers.

X7	X1	X2	X3	X4	X5	X6	X8	X9	X10	X11
Researchers-other Deptt. Communica tion	.315*	.221*	.308*	.148*	.168	.225*	.409*	.176	.257*	.387*

*Significant at .05 level of probability

Conclusion

To sum up, the study concluded that in the 'Agricultural Research System' which is a formal organization, intra-system communication patterns like peer communication, communication network, opinion leadership and communication with other departments occurred more among the members committed to research objectives. These interpersonal relations give rise to a feeling of belongingness and of collective interest which express themselves as 'morale'. These communication relationships can therefore, have a considerable influence, for good or for bad, on the general climate of the organisation.

The intra-system communication patterns of researchers were intricately related to individual and inter-system communication patterns. This may be fostered in an research organisation particularly dealing with agriculture, because the effectiveness of agricultural research organisation depends on mainly adequate communication among researchers, between researchers and extension personnel, and between researchers and the farmers. Among the research workers, both at national and international levels, effective communications are necessary in order to ensure that they are aware of the developments in their work and related fields, which have a bearing on their research work, as a source of new ideas and techniques and thus avoid unnecessary duplication. Between research and extension, they are needed in order to ensure that research findings reach the extension workers for transmission to the farmers, and also to ensure necessary feedback on problems with which the extension worker is faced in course of his work. Between researcher and the farmer, both direct and the indirect communication is essential.

REFERENCES

1. **ARNON, I.** (1968). Organisation and Administration of Agricultural Research. Elsevter Publishing Company Ltd., London.
2. BERRJEN, F.K. (1968). General and Social System. Rutgers University Press, New Brunsurck.
3. GLANZER, M. (1961), Techniques for the study of group structure and behaviour. Psychological Bulletin 58.
4. HIRCH, I. & MILWEL, W. (1958). Increasing the Productivity of Scientists. Harward Business Review. 36(2).
5. JAIN, N.C. (1970). Communication Patterns and Effectiveness of Professionals performing linking roles in a research dissemination organisation. Ph.D. Thesis, Michigan University, U.S.A.

6. LEAVITT, J.H. (1969). Some Effects of Certain Communication Patterns on Group Performance. Journal of Abnormal Psychology, 46.
7. LIONBERGER, H.F. & CHANG, H.G. (1970). Farm information for Modernising Agriculture: The Taiwan System. Preager Publishers, New York.
8. L1NGWOOD, DAVID. A. (1969). Communication Behaviour of Communication Researchers: Centre for Research on the Utilisation of Scientific Knowledge. University of Michigan, U.S.A.
9. MORENO, J. (1953). Sociometry in Group Relations. Council of Education Research, Washington, D.C.
10. PELZE, D.C. & ANDREWS, P.M. (1966). Scientists in Organisation. John Viley, New York.
11. SINGH, K.N. & PRASAD, N. (1972). A Study of Inter-and Intra-personal Departmental Communication. Indian Journal of Extension Education, New Delhi, 8 (1&2).
12. VICKERY, B.C. (1963). Scientific Information : Problems and Prospects. Minerva 2(1).

Chapter **8**

A Study of Communication Behaviour of Grass Root Extension Change Agents under Training and Visit System

M.P. Singh, Daya Ram

T&V system launched with the assistance of World Bank has reorganized Agricultural Extension Services in India. It infused flesh and blood into the field and thus with a added vigour and confidence it moved forward; state after state started signing the MOU with the World Bank. Consequently, almost all Indian States adopted this model.

Training and Visit System

The T & V is based on the Interpersonal Communication and diffusion of innovation theory. The innovations generated by the researchers are disseminated to the clients by extension agency over a period of time. The T&V system has in-built mechanism of transfer of technology (TOT), principles and guidelines for the selection of contact farmers on which this World Bank financed project is being operated. Generally, innovations are spread from their origin to users in a multi-step flow pattern. In a large system T&V, there are at least four steps through which information must flow (and hopefully in both directions) from researchers to the subject matter specialist, from them to field extension workers and through contact farmers to the majority of the ultimate users.

Objectives

1. To examine the information input pattern of the Extension personnel.

Methodology

Present study was conducted in three states i.e. Bihar, Madhya Pradesh and Uttar Pradesh during 2000-2001. For accuracy and success of project scientific methodology were followed. Selected Extension Personnel (EP) were interviews with the help of structured schedule. Besides interview, reports and relevant official publications as well as discussions with superior officers were also used as sources of additional information.

Sample and Sampling Procedure

For the present study, the field Extension Personnel (AEOs & VEWs) constituted the total population in the extension system. As the study is conducted in three states, out of each selected state regions having average performance were identified with the help of Directorate of Extension. Out of these selected regions one district was selected randomly. Within the district, all Sub-divisional Officer (SDO) Agriculture circles were covered and all field extension workers starting from AEOs to village extension workers (Kishan Sahayak) were interviewed. Besides, interview with these extension personnel, SDO (AE) were too interviewed informally to get relevant information which had bearing on the communication pattern of the extension system. Details about size are given below in table 1.

Table-1 : Number of Extension personnel covered under Study.

State	Direct	Coverage	
		AEO	VEW
Bihar	Bhabhua	28	148
M.P.	Rewa	31	154
U.P.	Gorakhpur	32	158
Total		92	460

Keeping in view the limitation of time and resources, it was difficult to interview the Extension Personnel at their place of working. Therefore, it was decided to interview them during fortnightly training. Thus, through correspondence, list of fortnightly training schedules were obtained and a team went on the scheduled date and time to conduct interviews. Each respondent was interviewed separately in order to avoid the peer and group influence.

Statistical Analysis

Associational relation between communication function and facilities were analyzed by means of zero order correlation. Multiple regression analysis was applied to find out the magnitude of effect of communication variables and antecedent variables both individually and in combination on communication patterns.

Path analysis adopting multivariate path model is suggested by Land (1969) was used to isolate the direct and indirect effects of exogenous variables on endogenous variables like information input, information processing and information output patterns.

Besides these statistical tools, student 't' test, standardization 'z' score and even percentage and average have been used in appropriate places and situations.

Definition & Operationalization of Concept

A number of terminologies have been used in this investigation with specific meanings and concepts. Operational definitions, according to Kerlinger (1965) is a specification of the activities of the researcher in measuring a variable or in manipulating it. Operational definition is a sort of manual of instructions to the investigators. The operational definition makes the meaning of a concept more explicit. Accordingly, various concept used in this study, have been operationalized, explaining their epistemic relationships and empirical measures:

Communication : Communication is the process of flow of information by which two or more people exchange ideas, facts, feelings, expression and signals so that each gains a common understanding of the meaning intent and application of the message. From the standpoint of the present study, communication is viewed as flow of information from extension system to client system.

Communication Behaviour: Communication behaviour of an individual, dved group or a social system is that which are systematic and exhibit some sort of regularity. The systematic communication behaviour of an individual may be acquisition, processing and dissemination of scientific messages. This may be communicated to peers, superior and personnel of other related organizations. In this study, only those messages were taken into consideration which were of worthy for farming community and transmitted to farmers particularly contact farmers.

Information Input Patterns (IIP)

Information input pattern refers to all the activities performed by an individual extension personnel for acquiring scientific and technical information related to agriculture, from various sources for performing his roles efficiently. The amount of information input was measured with the help of Information Input Index developed separately for them. The data were collected against each items and individual index was prepared for each extension personnel after assigning numerical scores to these items. Because these items had different ranges and units of measurement, the raw score of each item was converted into 'Z' score by adopting the following formula:

$$\text{Z-Score} = \frac{\text{Raw score - Mean}}{\text{Standard Deviation}}$$

Or

$$Z = \frac{X\text{-}X}{\text{SD or S}}$$

The Z - scores of all these items were summed up to obtain information input index of individual extension personnel.

Information Processing Pattern (IPP)

According to Thayer (1968), Information Processing deals with evaluation of received information; i.e. diagnosis, analysis, synthesis or deciding, storage of information i.e. noting, indexing, categorizing; and transformation i.e. reproducing in suitable form, amplification or reduction of initial information. Hence, the Information Processing refers to all the activities performed by an individual for synthesis, evolution, storage and transformation of scientific and technical information related to agriculture for performing his role.

The items and procedures of collecting data against each such item have been enlisted for constructinginformation processing index of extension personnel. Each item has numerical values, but as these items had different range and units of measurement, the raw score of each item was converted in 'Z' score applying the formula, discussed under informationinput index. The 'Z' score of all the items of index were added up separately to constantinformation processing index of individual extension personnel.

Information Output Pattern (IOP)

It refers to all the activities performed by an individual for dissemination of scientific and technological agricultural information.

The amount of information output for individual extension personnel was measured with the help of information output index following similar ways as in case of information input and processing phenomena.

Peer Communication (PC)

In many situations, one feels like sharing ideas and experience regarding technical and professional matter with the colleagues of his peer groups. Hence, the peer communication herein refers to the degree to which an individual extensionpersonnel communicate with his peers regarding technical and professional matters related to his specific role. Peer communicationhas also been measured by following same procedure mentioned above.

Extension Personnel- Research Communication (EP-RC)

Communication between Extension Personnel and with the researchers for acquiring relevant scientific information and to transmit farmer's problems to the researcher either at the time of fortnightly training or at his office are necessary for transfer of farm information and it's utilization by the farmers.

In the present study, EP-RC refers to the extent of communication pattern between EPs and Research for acquiring technical information and for providing feedback. There may be variation in the use of various sources, channels or in frequency of it's use for communicating with the researchers. This, in turn, leads to variation in the degree of communication between them. Almost similar case may be encountered in case of communicating farmer's problems.

For measuring the communication between EP and researchers, EP-RE index was prepared after converting the raw scores against each item, into 'Z' score and summating these scores.

Extension Personnel - Farmer Communication

Extension personnel communicate both farmers for disseminating farm information and to understand their farming needs and problems. The EP-FC refers to the degree to which an EP communicates with farmers about their need based agricultural innovation and solutions of their farm problems, through various methods and media.

The amount of EP-FC was measured by an index developed out of the responses against numbers of methods and media. Since each items has different range and units of measurement, the raw score of each item was transformed into 'Z' score. The 'Z'-score of all items were summer upto develop the EP-FC index of the individual extension personnel.

Extension Personnel - Farmer contact Spon (EP-FCS)

It refers to the degree to which an individual extension personnel contacts various types of farmers in connection with transmitting farm information. This was measured in terms of EP-FCS index. As the items were having different ranges and units of measurement, the score allotted to each of the item as converted into 'Z' scores of recommitting individual index.

Facility

Every communication act requires certain facilities. For example, in a formal organization, these facilities may be typing, cyclostyling, photography, artist, slide projector, overhead projector, movie projector, telephone printing etc. Likewise during the investigation, a list of various items as communication facilities were listed by asking the following questions from each of the respondents.

How many of the following facilities are available to you and to what extent these are made available to you for communication work? Please indicate among the following :

Each facility was read out by investigator and responses were recorded. 13 facilities were put as alternatives given in schedule under respective question.

The number of positive responses were accounted for each respondent in appropriate columns. One score was given for each item indicated.

The extent of availability like easily available, available with difficulty and not at all available were assigned 2, 1 and 0 scores simultaneously. The scores obtained for each respondents were converted into 'Z' score and then all such scores were added to construct facility index for each extension personnel.

Ownership of Conveyance

Respondents were grouped into three categories according to the mode of conveyance owned i.e. scooter/motorcycle, bicycle and no conveyance. Score 2.1 and 0 was assigned to the categories respectively.

Antecedent Variables

Antecedent variables, according to Rogers (1969) are those which proceed other variables, in order of time and which theoretically are expected to lead to or be followed by certain other variables. These variables may be situational or personal in nature.

Person and Shils (1965) stated that situation refers to all those things outside the organism to which actions may be related. Further, they mentioned that behaviour of any organism is called action when it is analyzed in terms of the situation in which it occurs.

Several researchers like Pelze and Andrews (1966), Winowood (1966). John (1971). Chakraborty (1971), Dubey (1977), revealed that certain situational and personality factors of researchers, extension personnel and farmers have association with their communication patterns of performance.

From this discussion, it was postulated that some selected situational and personal factors will be of great assistance to understand and predict the communication pattern of the extension personnel. Hence, some such variables were taken into consideration in this study. These variable are :

1. Age

It refers to the total years completed by an individual respondent at the time of interview. The respondents were asked to mention their age in terms of completed years. The data thus, revealed that the age of respondents ranged between 20 to 57 years. This range was divided into three groups, i.e. 20-33 years, 33-45 years and 45 & above. These groups were scored as 1, 2 and 3 respectively. Based on these scores standard scores were calculated.

2. Educational attainment

It refers to the formal education obtained from school to the university level by the individual respondent. The data pertaining to educational attainment were collected by asking their last and complete education. It was found that the level of education ranged from high school to postgraduate. These were categorized into four groups, i.e. High School, Intermediate, Graduate and Postgraduate and score 1, 2, 3 and 4 were assigned respectively to these categories.

3. Family Background

The family background of respondent was divided into two

categories, namely, urban background and rural background. Respondents having permanent affiliation with urban area (city/town) were categorized under urban background and those are from village setting, were grouped in rural background categories. All the respondents were found to have rural background.

4. Experience Gained

It refers to the period one has devoted to work on present position in the organization. It was measured by the number of completed years one has served on the present position. So far experienced is concerned it refers to the period one has devoted to work on present position in the present AEO/ VEW circle. One score was assigned to each year.

5. Training Received

Training means to educate a person to be fitted, qualified and proficient in doing some job. Training of extension personnel will be categorized into two groups namely- pre-service training and in-service training. Both trainings were considered jointly.

6. Job Satisfaction

It refers to the extent of satisfaction an individual draws from his present job. In this study, job satisfaction of the respondents was measured by self developed 3 points scale- satisfied, neutral and not satisfied. Score 3, 2, 1 was assigned respectively against each items (10 Nos). The average was worked out and taken as the job satisfaction score for each respondent.

Findings and Discussion

Profile provides a composite picture of the situation. In this case, the profile of respondents give an idea of their basic characteristics at a glance. The major categorization criteria used are age, education, experience and family background.

Data from table 2 reveals that majority of extension personnel (51%) were between 33-45 years of age followed by those of 20-33 years of age (34%) and 45 & above years of age (16.00) respectively. Thus, results clearly depict that majority of extension personnel belong to young age and adequate experience group. At present, drastic change in condition of farming population and agriculture production could be easily achieved through the planned strategies of human resource development.

Table-2 : Distribution of field Extension Personnel according to their Characteristic.

Age (yrs.)	20-30	187	34.00
	33-45	277	51.00
	45 and above	88	16.00
Education	High school	77	13.95
	Intermediate	129	23.39
	Graduation	235	42.56
	P.G. & above	111	20.10
Experience (yrs)	0-5	103	18.66
	5-10	209	37.86
	10-15	132	23.91
	15 & above	108	19.57
Family background	Rural	439	74.26
Training	0-5	283	51.27
	5-10	165	29.89
	10-15	104	18.84

As regards, the education, majority (42.56%) were graduates, followed by those having intermediate (23.39%) and postgraduate (20.10%) respectively. A significant number (45 out of 92) of AEOs were postgraduate.

So far as job experience is concerned, major part of extension personnel population (37.86%) were having 5-10 years of job experience followed by 10-15 years (23.91%) and 15 years and above (19.57%) respectively. It is clear from the analysis that above 82% extension personnel were having adequate working experience in their field of specialization.

Similarly majority of extension personnel (88.64%) belong to rural setting which no doubt is a benefiting indicator for this recognized extension system. But the low percentage of personnel with long training experience (18.24%) is a serious concern.

The overall profile-analysis of grass root extension personnel of these districts suitably fits with the nature of job being performed by the extension personnel. Without having sound farm of information and wisdom with knowledge and skill extension personnel could not be successful in their endeavour.

Individual communication Pattern

Individual communication pattern/behaviour of extension personnel have been studied in terms of information input, information processing and information output. Each of these patterns is described with average correlation, regression and direct and indirect effects.

Information Input Pattern of Grass root EPs

For information input, extension personnel utilize a number of sources and channels, against which data were collected primarily for computing information input indices. The same data have been used to workout the use extent of these sources and channels by the grass root EPs of the system.

Differential use extent of sources and channels

The use extent of each of these sources and channels among the grass root EPS (AEDs & VEWs) has been furnished in table 3. Based on their ranks in terms of use extent, weighed scores were worked out for each of the sources/channels. These scores have been used to reveal its overall use extent by the field extension personnel.

Agriculture Extension Officer (AEOs) use all the sources and channels for information input. However, the use extent scores being less than one with respect to research publication, professional affiliation. Training camps.

specialists lecture, consulting scientists, visit to experimental/ research farms, national demonstration, and farm radio broadcast is precisely indicative of its scare use by them. The remaining three sources, namely leaflets and bulletins, departmental meetings, and popular extension journals are the three most used sources by the BLEP.

The village extension workers (VEWs) utilize all the sources and channels except research publication and professional affiliation, But departmental meeting is the only source having use-extent score more than one. It is, thus, amply clear that the rest of the sources and channels are rarely used by them.

Table-3 : Average use-extent of different sources and channels for information input by extension personnel.

Source/Channel scores	BLEP	VLEP	Weighed
Departmental meetings	1.92	1.15	19
Leaflets & bulletins	2.57	0.80	18
Popular extension journals	1.89	0.82	17
Research publications	0.00	0.00	0
Professional affiliations	0.15	0.000	2
Training camps	0.80	0.50	12
Specialist lectures	0.68	0.05	6
Consultation with scientists	0.37	0.15	6
Visit to experimental/ research farms	0.85	0.10	10
National demonstrations	0.65	0.45	24
Farm radio broadcasts	0.74	0.57	12

Based on the overall use extent scores, all the eleven sources and channels have been categorized in three groups as presented in table 4.

Table-4 : Categorization of Source and Channel based on weighed scores

Most used	Moderately used	Least used
Departmental meeting	Training camps	Farm radio broadcast
Popular extension journals	Specialist lecture	Research publication
Leaflets and bulletin	National demonstration	Professional affiliation
Consultation with scientists visit to experimental/ research farm		

It is obvious that the most used sources and channels by the extension system as a whole for the acquisition of scientific and technical information were departmental meetings, popular extension journals, leaflets and bulletins, visit of farm, radio broadcast; research publications and professional affiliations are the least used ones.

Information input Amount

The amount of information input for each field extension personnel has been measured in terms of information input index. The input indices of the field EPs range from 0.86-33.18. The overall average information input of these extension personnel comes to be

5.33. The average input indices of these extension personnel shown in table 5.

Table-5 : Average input indices of EPs and significance of their difference.

Extension personnel	Average	't' value
AEO	10.36	5.36**
VEW	4.35	

*Significance at 0.01 level.

Obviously depicts that there is a wide range of variation in the information'input amount among the AEOs and VEWs. Information input of VEWs is less than the overall average. Test of significance of difference between AEDs and VEWs are highly significant.

Correlation and Regression of Communication on Variable with Information Input

The correlation coefficient of information input with the other communication variables have been furnished in table 5. Out of seven communication variables, six are found to have highly significant correlation with input. These are information processing. Information output, EP-RC, PC, EP-FC, EP-FCS and facilities.

Table-6 : Correlation and Regression of communication variables with information input Indices.

Independent variable	Processing	Output	EP-RC	PC	EP-FC	EP-FCS	Facility
Correlation coefficient	.658**	.766**	.814**	.547**	.526**	-163	.621
Regression coefficient	.373	.212*	.208*	.341	.046	-025	.116*

* Singificance at 0.05 level of probability

** Significance at 0.01 level of probability

$R^2 = 0.7892$

$F = 43.33^{**}$

Multiple analysis was carried out with the seven communication behaviour variables. The obtained partial regression coefficient and the value-of R^2 have also been furnished in above table 6. The significance partial regression coefficients of output, EP-RC, and the facility indicate that these three variables exert their influence on information input index. However, it can be concluded from the value of R2 0.7892J that all seven variables explain 78.92 percent variation in the information input of EPs.

Correlation and Regression of Communication and Antecedent Variables with Information input

In a search for the other variables which can account for the rest of 21.08 per cent variation in the information input, antecedent variables were added to the communication behaviour variables, and multiple regression analysis was again carried out. The value of correlation coefficient between input and antecedent variables are furnished in table 7. Information input indices exhibits significantly positive correlation with all the antecedent variables except family background.

Table-7 : Correlation between Information input and Antecedent Variables.

Variables	Age	Education	Service	Training	Family back-ground	Job satisfac.
Correlation	.231*	.631	.371**	.582**	-221*	.345**

* Significance at 0.5 level of probability

** Significance at 0.1 level of probability

$R^2 = 0.8143$

$F = 23.12^{**}$

Family background shows negative correlation with input indices. The value of R^2 (0.8143) indicates that addition of antecedent variables to communication variables can explain further variation in the input only to the tune of 2.51 per cent. Looking into the very small contribution of antecedent variables, further analysis was carried out only with the communication variation.

Direct and Indirect Effects

The multivariate path analysis is used to explain direct and indirect effects of exogenous variables on endogenous variable. Here the input index is regarded as endogenous and to each seven communication variables as exogenous variables.

Table-8 : Direct and Indirect effect of Exogenous Variables on information input index as Endogenous Variables based on path Coefficient.

Exogenous variables	Direct effect	Total indirect effect	Largest effect	Indirect through
Processing	.172	0.482	.589	EP-RC
Output	-.625	1.391	.685	PC
EP-RC	.582	0.236	.591	PC
Feer	-.074	0.261	.827	EP-FC
Communication EP-FC	.053	0.475	.467	EP-RC
EP-FCS	-.009	-0157	.052	EP-RC
Facility	.133	0.448	.486	EP-RC

The standardization path coefficient as furnished in table 8 are directly comparable. EP-researcher communication, processing and facility are the first three exogenous variables which exerts maximum direct effects on input amount of EPs in descending order. The remaining exogenous variables have comparatively very small direct effects.

The total indirect effects of exogenous variables indicate that output, peer communication, processing, EP-FC, and facility exerts respectively first, second, third, fourth and fifth largest total indirect effect as against their seventh, sixth, second, fourth and third direct effects. It is further revealed that EP-RC exerts maximum indirect effect on input through other variables, closely followed by peer communication.

DISCUSSION

Departmental meetings have the highest overall use extent scores. It suggests that the departmental meetings which are held regularly to discuss the achievements and failure therein are also most commonly used for disseminating new scientific and technical information to the extension personnel. The findings regarding the extension publications, training camps, farm radio broadcast and visit to experimental research farms being among the most used sources and channel for information input. The greater degree of euthenics and easy approachability might be the two possible reasons for this. Moreover, the publications being printed materials have the less possibility of distortion or loss of information, so this adds to its credibility. Besides, publication is supplied to the EPs by the state department of agriculture or by the other sources, mostly free of cost. Thus, they are easy to obtain and are regular, handy and less costly. They also provide readymade reference material as and when required. These might be some probable reasons for extension publications being one among the most used information input sources among the EPs. Training camps and farm radio broadcasts are other frequently used sources of information. As fortnightly trainings are integral part of T&V system, fields extension person get information about new innovations at the same time when subject matter specialist provides the solution to their field problems. In addition to this SMS, experts of other concerned organization (public and private) too attend the fortnightly meeting and transmit information about the new innovations in agricultural sector. Likewise, radio broadcasts several programmes on agriculture in which researchers convey the relevant messages on every facet of farming. Radio being available at cheapest rates, it is not beyond the reach of extension personnel. Visit to experimental farms is also a most used source of information and provides an opportunity to observe things personally and to understand, discuss and ask the scientists for clarification.

Sources like national demonstration, specialist's lectures and consultation with scientists are moderately used sources because field extension personnel seldomly get an opportunity to participate and visit. Besides, lecturers of experts are rarely organized at SDO (AE) level.

Research publications arid professional affiliation are the least used sources of information input. Probably being research oriented, they are not of much practical significance in the actual field situation for the EPs. Moreover, the scientific languages, used is also such, that extension personnel only with post graduate and higher degree would easily comprehend it and will have interest therein.

Table 5 shows that the VEWs have the comparatively low information input average than the AEOs. Even information input average of AEOs can not be said satisfactory. This is probably because of deficient of facilities for in service training to VEWs and due to inadequate arrangements for keeping them well abreast with the latest research findings. It has obvious implications. It is accepted by all that VEWs are the key linking force having great potentiality to multiply farm information utility many folds, but can not be expected of them when they themselves do not have adequate relevant information. Such a state of affair is neither desirable nor tenable, particularly when the agricultural technology has gained a new height of its achievement. Hence, it is essential that they be provided with the latest farm information to add to their efficiency and creditability.

It is further indicated that input has significantly positive correlation with six variables. EP-FCS does not show significant correlation with input. Multiple correlation coefficient of input with communication variables indicates that all the seven communication variables under study, can account for 78.92% variation in input indices. In an attempt to identify other variables explaining rest of 21.08% variation in input indices, antecedent variables were added to communication variables and again multiple regression analysis of input with all these combined variables was made. Multiple correlation coefficient thus obtained, indicates that the antecedent variables did not contribute much in explaining input indices. Probably the communication variables themselves take care of all these antecedent variables. The path analysis leads further to conclude that out of the seven exogenous variables, EP-RC, processing and facility have maximum direct effects on the information input amount of the EPs whereas output peer communication, processing, EP-FC and facility exerts maximum total indirect effects on it.

CONCLUSION

The success of any extension system depends mostly on the communicability of the functionaries working in the organization. The present study was undertaken necessarily to investigate the communication behaviour of field extension worker's (AEOs and VEWs) under T&V system. The period covered was from 9 May 1994 to 9 July 1995. The data was collected from three districts Bbabhua, Rewa and Gorakhpur one each from Bihar, Madhya Pradesh and Uttar Pradesh respectively by covering all the sub divisional agricultural extension officer circle. Altogether 92 agricultural extension officers (AEOs) and 460 village Extension workers (VEWs) were interviewed by employing structured schedule developed for the study. Besides interviews, the informal discussions with senior officials, observations of investigators were made to utilize it is drawing inferences while explaining communication processes. Findings of the study are as follows :

1. Majority of the field extension personnel were of rural background (74.26%) in 33-45 years of age group (51%) with

graduation as educational qualification (42.56%), 5-10 years of working experience (37.46%) and attended maximum 5 trainings (excluding fort nightly trainings) (51.27%).

2. Leaflets and bulletins, departmental meetings and popular extension journals are most used channel of information by AEOs while VEWs largely dependent on departmental meetings only.
3. Farm broadcast, research publications and professional affiliation are used by field extension worker for getting scientific and innovative technical information.
4. The average amount of information input was found higher in AEOs as compared to VEWs whose information input amount was found below the average.
5. The information input has significantly positive correlation with processing, output, EP-RC, Peer communication and EP-FC.
6. Likewise, antecedent variable age, education, service experience training and job satisfaction had significantly positive correlation with input but family background has significantly negative correlation.
7. The multiple correlation coefficients between input and other communication variables revealed that all these account for 78.92% variation in the input amount of Grass Root extension personnel.
8. Antecedent variables associated with communication behaviour explained only 81.43% variation in the input indices which obviously revealed that these antecedent variables contribute only 2.51% in explaining the information input behaviour.
9. EP-RC, processing, facility and EP-FC exerted maximum direct effect on information input amount of extension personnel.
10. Problem regarding input non-availability was mostly raised by the field extension worker during fortnightly training but this too happened often.
11. So far perceived problems are concerned least opportunity to outside visits and to interact with scientist directly were revealed by most of the extension workers. Heavily reliance on oral communication and tackling problems in adaptation of any practices were perceived by FEWs as some another important problems which dealing with organizational objectives.

Chapter 9

Interrelationship among Demographic Characteristics (Variables), Media Availability, Media Utilization and Development Message Needs

Upayana Singh, O. P. Mishra, Vijay Avinashilingam

A country like India, where a major proportion of population i.e. 74.30 percent (India, 2002) resides in rural areas, the role of communication in the development cannot be ignored. When we talk about development of India, the mind automatically switches to the development of rural areas. The present structure and practices of the well-established information sources viz. Radio, Television and Newspaper are not quite conducive to bring rural areas into communication network. The reason is lack of consideration given to the rural peoples. Effective communication of development message, no doubt, is key to success. The development messages needs varies from person to person. Similarily the type of media available and their utilization pattern also varies. This variation is due to demographic characteristics. So it is necessary to understand the relationship between demographic characteristics message needs. The media utilization pattern and development message needs. This will help the planners in formulating information dissemination strategy and in developing need-based information package. Keeping above discussion in mind, the present study was conducted to study the relationship of demographic characteristics with development message needs, availability and utilization pattern of information sources.

Research Methodology

A sample of 105 farm families was selected from two purposively selected villages of Rudrapur block of Udham Singh Nagar district of Uttaranchal. The head of each family was interviewed. Media availability has been operationalised as the possession of media such as Radio, Television, Newspaper and magazine by the respondent. Media utilization has been operationlised as periodicity of utilization of different media namely Television, Radio, Magazine and Newspaper i.e. regularly, occasionally and never. In addition to this purpose, amount of time daily spent and segment of day utilized are also studied. Development messages-need has operationlised as information related to all public service areas and national development sectors such as agriculture, health, family planning, nutrition, formal and non-formal education. Following were taken as the demographic characteristics of farm families :

Average age of family, total education status, average education status, socio-economic status, estimated annual income (Rs '000s), Land holding (Bigha), Material possession, Size of the family, Type of family and Caste.

The data was collected through semi-structure (Open-ended) interview schedule which was further supported by other methods such as observation, informal interpersonal discussion and PRA techniques.

Results and Discussions

Interrelationship among Demographic characteristics (variables), Media availability, Media utilization and Development Message Needs Interrelationship among different variables were studied by working out the correlations. The selected variables with their codes have been presented in Table 1 and these codes have been subsequently used in the inter correlation of the variables presented in Table 2. A cursory look at the Table 2 make it clear that even the demographics (profile) variables were related with each other. Total educational status of the family was positively and significantly correlated with socio-economic status, income, land, materials possession, size of family, type of family and caste at one percent level of significance. This shows that farm families who were high on total educational status score. Since education (i.e. formal education) requires investment of money and these variables are also indicative of monetary status of the family. This may be a reason for significant and positive correlation among the variables. However, it also indicates that respondents having belonged to upper strata of the caste hierarchy were also high on educational status score.

Socio-economic status was positively and significantly related with income, land, material possession, family size and caste at one percent level of significance, whereas at five percent level of significance, it was also positively and significantly related with type of family.

Income was positively and significantly related with land, material possession, size of family and type of family at one percent level of significance. It was also found positively and significantly correlated with caste at five percent level of significance.

At one percent level of significance land holding was positively and significantly correlated with material possession and family size whereas at five percent level of significance, it was positively related with type of family and caste.

Material possession was positively and significantly correlated with family size, type of family and caste at one percent level of significance.

Size of family was positively and significantly correlated with type of family and caste at one percent level of significance. This indicates that respondents having joint family and belong to general caste are also having larger number of member in the family.

The finding in the study generally support the hypothesis that education, income, land holding and material possession have influence upon the socio-economic status of the people.

Relationship Between Demographic (Independent) Variables and Media Availability

Availability of media to respondents was positively and significantly correlated with their education status, socio-economic status, income, land holding and material possession at one percent level. However, positive and significantcorrelation was also observed with family size and caste at five percent level. This means that families with high education score and from high-income group had more number of mass media available with them. This is well known fact that for reading print media, education is required and for purchasing different kinds of mass media, money is required. The present study also support this fact to a great extent.

Relationship Between Independent Variables and Media Utilization Pattern

Table 2 reveals that five independent variables such as education, socio-economic status, income, land and material possession were positively and significantly correlated with media utilization at one percent level of significance. Family size and caste were found to be positively and significantly correlated with media utilization at five percent level of significance. However, non-significant and positive correlation was also observed with type of family, and average age of the family. The findings of the study reveals that farm families, who were high on total education status score, socio-economic status, income, land holding and material possession were also high on mass media Utilization score.

Table-1 : Selected variables with their codes

Variables codes	Variables
X1	Average age of family
X2	Total education status
X3	Average education status
X4	Socio-economic status
X5	Estimated annual income (Rs 000s)
X6	Land holding (bigha)
X7	Material possession
X8	Size of family
X9	Type of family
X10	Caste
X11	Media availability
X12	Media utilization pattern
X13	Agricultural information needs
X14	Health information needs
X15	Family planning information needs
X16	Education information needs
X17	Nutrition information needs
X18	Development message needs

This means that mass media were most utilized by higher socio-economic status group. Results of this study are similar to the findings reported by Gaikwad et al. (1973), Kukerty and Singh (1994); Swarnkar and Aggarwal (1999).

Relationship Between Media Availability and Media Utilization

When we analyze the correlation between media availability and media utilization. Table further reveals that there was positive and significant relationship between these two variables at one percent level of significance. We can draw the meaning from this result that people who had more number of media available with them utilized

Table-12. Interrelationship among demographic characteristics (independent variables), media availability, media utilization pattern and development message needs

	X1	X2	X3	X4	X5	X6	X7	X8	X9	X10	X11	X12	X13	X14	X15	X16	X17
X1																	
X2	-0.10																
X3	0.06	0.06															
X4	-0.12	0.64**	=0.01														
X5	-0.06	0.75**	0.001	0.89**													
X6	-0.05	0.57**	-0.03	0.91**	0.86**												
X7	-0.08	0.71**	0.02	0.86**	0.84**	0.76**											
X8	-0.25*	0.76**	-0.05	0.59*	0.66**	0.53**	0.59**										
X9	0.09	0.28**	-0.11	0.25*	0.25**	0.22*	0.30**	0.46**									
X10	-0.16	0.25**	0.09	0.30**	0.25*	0.20*	0.29**	0.28**	0.12								
X11	0.05	0.41**	0.00	0.42**	0.43**	0.41**	0.53**	0.20*	0.08	0.22*							
X12	0.03	0.29**	-0.05	0.37**	0.37**	0.36**	0.45**	0.22**	0.14	0.23*	0.77**						
X13	-0.03	0.15	0.04	-0.05	-0.16	-0.11	-0.11	0.00	-0.09	-0.10	-0.22*	-0.13					
X14	-0.15	-0.03	0.00	-0.13	-0.07	-0.16	-0.11	0.08	0.07	-0.01	-0.14	-0.12	0.17				
X15	0.16	-0.17	-0.09	0.01	-0.02	0.04	0.05	-0.01	-0.06	-0.13	-0.02	-0.10	0.08	0.18			
X16	0.14	0.14	-0.06	0.04	0.13	0.04	0.04	0.17	0.16	-0.12	-0.06	0.01	0.07	0.25*	0.04		
X17	-0.19*	0.10	0.03	0.04	0.08	0.00	0.02	0.07	0.01	0.00	-0.01	-0.05	0.10	0.39**	-0.02	0.21*	
X18	-0.05	0.00	-0.01	-0.04	-0.01	-0.08	-0.04	0.11	0.03	-0.12	-0.16	-0.13	0.51**	0.72**	0.36**	0.55**	0.65**

*Sinificant at 5% level

**Significant at 1% level

df = 103

the media most. It is interesting to note that independent variables which were related with media availability were also related with media utilization.

Relationship Between Demographic (Independent) Variables and Development Messages Needs

Information need of farm families related to agriculture, health, family planning education and nutrition were worked out. Development message needs were calculated by adding number of needs found in all the five areas as mentioned above. Relationship of development messages needs was seen with independent variables, media availability and media utilization. Inter-correlations among all the five areas of development messages needs with different areas of information needs were also worked out. The correlation values have been presented in Table 2. Table 2 reveals that development message need was positively but non-significantly related with independent (Profile/demographic) variables like, total education score of the family, size of family and type of family. However, negative but non-significant correlation was found with average age of the family, socio-economic status, income, land, material possession and caste. Findings of the study indicate that independent variables had no influence on development message needs of the farm families. This aspect need to be further examined with larger sample size.

It is further reveals by the Table 2 that media availability and media utilization pattern were negatively but non-significantly related with development messages needs. This means that media availability and media utilization have no impact on development message needs.

Table 2 indicates that all the five areas of needs viz. agricultural needs, health needs, family planning needs, education needs, and nutrition needs were positively and significantly correlation with development messages needs at one percent level of significance. When inter-correlations among all the five areas of needs were examined, it was found that health need was positively and significantly related with nutrition need at one percent level of significance. Health needs and education needs were positively and significantly related with each other at five percent level of significance. Education need was positively and significantly related with nutrition need at five percent level of significance. The findings indicate that people who have higher need in the areas of health have also higher level of needs in the areas of education and nutrition.

People who were high health require good nutrition and education breeds consciousness about health and nutrition among people.

The study further reveals that nutrition needs of the farm families was negatively and significantly related with the average age of the family. Media availability was negatively and significantly related with agriculture needs. We can draw meaning from these two findings that families having higher average age require less information related to nutrition. This may be due to the reason that families from higher average age group have experienced an old age person to guide about the nutrition. Further, negative correlation between media availability and agricultural need may be due to the reason that families having more number of media available with them might be receiving agricultural information from different media. Thus, they may have less number of information need related to agriculture.

Conclusion

Year after year the urge to come up in every sphere of life human beings are always in the increasing trend. Different societies need different information to face the life competitively. The relation of development news from rural areas and communication of the same to them requires as special understanding of interrelationship between demographic characteristics, development messages needs, media availability and media utilization pattern.

This study provides a foundation for people to answer the existing question of their changing message needs. Finally, it can be stated that the study would be definitely useful to scientists, extension workers, development communicators, press etc., engaged in designing, development and disseminating development messages to the rural populace.

References

1. Gaikwad, V. R.; Tripathi, B. L. and Haque, S.M.S. (1973). Rural Social Structure and Communication in an Indian Village Center for Management in Agriculture, IIM Ahmedabad (Cited by Kumar, 2000).
2. India 2002, A reference manual. Ministry of Information and Broadcasting. Govt. of India, New Delhi.
3. Kukrety, N. and Singh, B. B. (1994). Mass-media and Interpersonal Channels for Agricultural Information in the Hills. Agricultural Ext. Review, 6(1) : 28-29.
4. Kumar, M. (2000). Communication of development messages among

rural poor-An ethnographic study. Thesis, M.Sc., G.B. Pant University of Agril. & Tech., Pantnagar.

5. Swarnkar, V. K. and Agarwal.D. (1999). Communication behaviour of farmers in accepting new farm technology under watershed project areas. Maharashtra Journal of Extension Education, Vol. 17 : 42-44.

Chapter 10

Communication Behaviour of Farmers in Training and Visit Extension System

S. N. Pandey

Activities performed by an individual for evaluation, transformation and preservation of the acquired information within his framework of references reflect the information processing behaviour. The extent of information processing at the consumers' level is an important indicator of the source trustworthiness, the message usefulness and the effectiveness of the communication channels. The study attempts to analyse various components of communication behaviour and thereby offers a logical explanation of patterns found in communication behaviour of the farmers.

Communication behaviour as a composite variable consists of three main components i.e. information processing and information output behaviour. Information input behaviour is mainly conceptualized as the activities performed by an individual for acquisition of relevant information from its various sources. In fact, the message-consumers in a communication process do not, out rightly, accept the information conveyed to them, rather, they undergo for rigorous mental and physical exercises for examining the pros and cons of the communicated information in many ways. There is also an inverse relationship between information processing and it's rate of acceptance and adoption. Therefore, reduced information processing at the consumers' level is as an important indicator of the source trustworthiness, the message-usefulness and the effectiveness of the communication-channels. The behavioural sub-components of information output of farmers are their knowledge, attitude towards the communication and the messages, adoption, feedback and interpersonal communication of the messages for relaying them to fellowmen. Therefore, the analysis of communication behaviour of farmers would provide valuable information about communication effectiveness in the training and visit extension system which was the World Bank Project, is begin widely introduced in the developing countries. It is needless to reiterate that the T and V system is a technically dominant system and communication of relevant farm production messages at a faster rate is vital for its success.

Methodology :

Location and sample :

The study was conducted in Chambal Command Area Development Project of Kota in Rajasthan in 1979. The project area was divided into three districts, namely Sultanpur, Baran and Bundi and out of them Bundi district was selected on random basis for the study. 20 villages were randomly selected, 10 from each of the two development blocks, namely Taleraand Keshoraipatan in the selected district. Again, out of the 20 selected villages, 100 contact and 100 non-contact farmers were randomly selected at the rate of five in each village.

Empirical measurement of concepts :

Information input behaviour :

It was empirically defined as the activities performed by an individual farmers for acquiring scientific farm information from different sources. The information sources were categorized as personal localite, personal cosmopolite, and impersonal and mass media sources. The use extent of different farm information sources and the information input amount were measured with the help of information input index. Each item in the index was arbitrarily assigned a numerical score. Since the units of measurements as well as the range of arbitrary scores assigned to different items in the input-index varied, the scoring patterns were made uniform by assuming the maximum 100 scores to be assigned to each item. The real score for each item was obtained by adopting the following formula :

$$\text{Real score of an item} = \frac{\text{Arbitrary score assigned to the item}}{\text{Maximum possible score}} \times 100$$

Weightage were also assigned to each of the items in the input index in terms of their importance in the T and V extension system. The computed real score was multiple by the respective weightage of the items. Then, the sum of the real scores in the input-index was divided by the sum of the weights. The final figure obtained, thus, was reckoned to be the information-input amount of an individual farmer.

Information processing behaviour :

The use-extent of various evaluation and preservation items and the information processing amount of the individual farmer was measured by developing an information processing index. Each item of the index was assigned a numerical score. Since the items had different ranges and units of measurement, the arbitrary scores assigned to them were converted into real scores by adopting the procedures as described in case of information input-index. The sum of the items which was taken to be the information processing amount of an individual farmer.

Information output behaviour :

Information output behaviour of an individual farmer referred to his level of knowledge, attitude, use-adoption, feedback and the extent to which he conveyed the farm messages to his fellow farmers. Like information input and processing indices, the output-index also contained several items which were assigned scores arbitrarily. The information output amount of an individual farmer was computed following the procedures as discussed in case of information input behaviour.

Findings :

Comparative use-extent of sources and channels for information inputs :

It is evident from the Table-1 that none of the contact as well as non-contact farmers used the information sources such as farm scientists, campaign, visit to research centers, farmers' training and newspaper. All mostly interacted with personal localite sources (neighbours, friends, relatives etc.) and VLWs for their farm information.

A high percent contact farmers used the sources like Village Extension Workers (97.9%), neighbouring farmers (85.5%), relatives (84.4%), family members (84.2%) and radio (72.6%). The average contact farmers received farm information from Agricultural Extension Officers (AEOs) followed by minikit trails (44.2%). Farm

Table-1. Use-extent of farm information sources and channels and percentage of farmers using them.

Information sources/channels	Use-extent score (%)		Overall score (N=200)	Percentage of farmers	
	Contact farmers (n1=100)	Non-contact farmers (n2=100)		Contact farmers	Non-contact farmers
Personal localite					
Family members	67.4	74.5	40.9	84.2	93.9
Neighbours and friends	95.5	95.6	95.5	100.0	100.0
Farmers of other villages	42.3	30.6	36.4	89.5	68.7
Personal cosmopolite					
Village extension worker	64.9	60.1	62.5	97.9	89.3
Agricultural extension officer	21.3	11.8	16.5	53.0	52.6
District extension officer	10.8	8.0	9.4	30.5	22.6
Subject matter specialist	8.3	3.4	5.8	24.3	10.4
Farm scientist	0.0	0.0	0.0	0.0	0.0
Officials of private agro-centers, cooperatives and banks	7.3	0.2	4.6	22.1	5.2
Impersonal					
Minikit trails	23.9	8.4	16.1	44.2	12.1
Field days	7.9	6.5	4.2	10.5	7.8
Agro-exhibition	21.0	17.4	19.2	23.1	15.6
Farmer's training	0.0	0.0	0.0	0.0	0.0
Crop campaign	0.0	0.0	0.0	0.0	0.0
Visit to experimental centers	0.0	0.0	0.0	0.0	0.0
Mass media					
Radio	28.6	22.8	25.7	72.6	54.8
Farm journals/magazines	0.5	0.4	0.4	0.9	1.7
Farm bulletins/leaflets	5.1	1.6	3.3	9.5	6.9
Handouts	6.1	2.9	4.5	22.1	7.8
Agro-films	21.0	20.4	20.7	33.7	36.5

*The sum of percentage is more than 100 due to multiple responses.

literature and field days were used by a few contact farmers. The similar pattern of information-source-use observed at the level of non-contact farmers.

On the basis of pooled use-extent, farm information sources have been summarized in Table-2.

Table-2. Categorisation of sources and channels based on overall use-extent scores.

	Most used sources/ channels	Moderately used sources/channels	Least used sources/channels	Not at all used
1.	Neighbours and friends	Relatives	Radio	Farm scientists
2.	Family members	Farmers of other villages	Agro-films	Farmers trailing
3.	Village extension workers		Agro-exhibitions	Crop campaign
4.			Agril. Extension officer	Visit to experimental centers
5.			Minikit trails	Newspapers
6.			District extension officer	
7.			Field days	
8.			Subject matter specialists	
9.			Officials of private agro centers, cooperatives and banks	
10.			Handouts	
11.			Farm bulletins/leaflets	
12.			Farm journals/magazine s	

Information input amount :

The input indices of farmers ranged from 0 to 100. The average information input index of contact farmers (14%) for personal cosmopolite, impersonal and mass media sources of information was significantly higher than the non-contact farmers. However, the average input index of the contact farmers for personal localite sources did not differ significantly from that of the non-contact farmers (43.2%). The overall input index of the farmers was found to be 12.6% which indicated that farmers, in general, were poorly exposed to various sources of farm information.

Correlation-regression of variables with information input indices of farmers :

Out of the 20 independent variables, only three i.e., production facilities, innovation proneness, and conservatism liberalism were significantly associated with the input indices for personal localite information sources. The variables such as age, farm size, education, occupation, caste, material possession, social participation, socio-economic status, cooperativeness, fatalism scientism, authoritarianism, economic motivation, affiliation motivation, compositeness, information processing and information output could not establish significant association.

The information input indices for personal cosmopolite, impersonal and mass media sources of farm information were significantly associated with education, caste, material possession, social participation, socio-economic status, production facilities, innovation proneness, co-operativeness, economic motivation, information output, fatalism-scientism and conservatism-liberalism. The association with conservatism and fatalism was only found to be negative.

The multiple regression analysis indicated that production facilities, innovation-proneness and conservatism expressed significant influence on input indices for personal localite sources, the similar analysis in case of information input for personal cosmopolite, impersonal and mass media sources of information revealed that only two variable i.e., production facilities and information output exerted significant influence over it.

Innovation-evaluation :

Percentage of farmers observing different evaluation items of farm innovation has been presented in Table-3.

It is evident from the Table that the farmers mostly considered economic profitability (94.28%), production resources available to them (93.33%) and simplicity-complexity (92.85%), witnessed demonstration on the plots laid by neighbouring farmers (87.14%), conducted trails (73.5%), discussed with the fellow farmers (63.33%) and the extension workers (54.76%) before finally taking a decision to adopt farm innovation. Accepting the farm innovation as unreserved and consideration of taste were least prevalent. The evaluation like discussion with family members and relatives and susceptibility to pests and diseases were considerably used.

Table-3. Percentage of farmers observing different items of farm innovation evaluation*.

Evaluation itmes	Contact farmers (N1 = 95)	Non-contact farmers (N2 = 115)
Accepting the innovation as unreserved	1.01	0.1
Discussing with family members	37.90	40.0
Discussing with fellow members	60.00	66.1
Discussing with relatives	41.00	36.5
Discussing with village extension workers	57.90	52.2
Discussing with other extension workers	10.50	5.2
Discussing with farm specialists	8.70	3.2
Considering simplicity-complexity in use	89.50	95.6
Considering availability of production resources	95.80	91.3
Considering economic profitability	96.80	92.2
Considering pest and disease infestation	25.30	42.6
Witnessing the results of demonstrations fields laid by fellow farmers	84.20	89.6
Considering taste	0.00	6.1
Conducting trails	71.50	75.5

*The sum of percentage is more than 100 due to multiple responses.

Innovation-preservation :

For the innovation preservation, the contact as well as the non-contact farmers mainly depended on memory. None of them was found to note the new farm information in specified note-book. A few farmers were found noting the farm information in a common note-book (usually a diary).

Innovation processing amount :

The contact and the non-contact farmers did not differ significantly with respect to their innovation processing indices. This indicated that they adopted the similar pattern of information processing.

Correlation-regression of variables with information processing :

Information processing was significantly and positively correlated with conservatism-liberalism and cooperativeness. It showed significantly negative association with innovation proneness and information output. But its association could not be significantly established with age, farm size, education, occupation, caste, material possession, social participation, socio-economic status, production facilities, fatalism-scientism, authoritarianism, economic motivation, affiliation motivation, cosmopoliteness and information input. The multiple regression analysis revealed the innovation proneness and information output exerted significantly negative influence on information processing whereas, economic motivation and conservatism-liberalism exerted significantly positive influence.

Information output behaviour

The findings indicated that the total mean knowledge indices of the contact and the non-contact farmers about scientific rice production messages were 50.15 and 45.81 percent, respectively. It means there were 49.85 and 54.19 percent knowledge gap at the levels of contact and non-contact farmers, respectively. The mean attitude scores of the contact farmers (2.58) and the non-contact farmers (2.43) were below the neutral score (3.0). It could be, thus, inferred that the farmers, in general, held unfavourable attitude towards the T and V extension system. However, the mean attitude score of the contact (3.51) as well as the non-contact farmers (3.20) indicated favourable attitude towards the scientific rice production innovations. The findings further indicated that the contact and non-contact farmers adopted the scientific rice production practices upto 52.0 and 48.0 percent respectively, the difference being non-significant. As regards feedback, none of the farmers made feedback with the research system. It was confined to the extension system only and it was also non-purposive. The feedback analysis further revealed that hardly 16.9 percent of the non-contact farmers made feedback with the extension system which was significantly lower than the contact farmers. The interpersonal communication indices (the extent to which an individual farmers relayed the new farm messages to his fellow farmers or the fellow farmers interacted with him for seeking farm advise of the contact and the non-contact farmers) were 67.56 and 55.10 percent, respectively and the difference was significant. This indicate that a considerable number of contact farmers were not as communicators.

Information output amount :

The output indices of the farmers ranged from 27.0 to 83.0 percent.

However, the average output of the client system reckoned to be 50.0 percent. The average output amount of the contact farmers (55.0%) was found to be significantly higher than that of the non-contact farmers (45.0%).

Correlation-regression of variables with information output behaviour :

Information output indices of farmers showed significantly positive association with farm size, education, caste, material possession, social participation, socio-economic status, production facilities, innovation-proneness, cooperativeness, economic motivation and cosmopoliteness. However, its correlation with fatalism-scientism, conservatism-liberatism and information processing was significantly negative. The regression analysis indicated that innovation-proneness, economic motivation, cosmopoliteness and information input (cosmopolite information sources) exerted significantly positive influence on the information output. However, authoritarianism and processing exerted significantly negative influence.

Discussion and suggestions :

The overall picture of the information source utilization of farmers indicates that the contact and non-contact farmers mainly depend on the personal localite sources i.e. the neighbours and friends, family members and also on the village extension workers for their farm information inputs. The important field extension functionaries such as Agriculture Extension officers, district and sub-divisional agriculture extension officer, subject matter specialists and also a few important extension methods/media like demonstration, minikit trails, agro-films etc. were mentioned by a few farmers only. This indicated that these sources could not be gainfully utilized by the farmers for their information inputs. The T and V system originally envisaged that the VEWs, AEOs, DAEOs and the district level SMSs should devote most of their time in field visits. Even the senior SMSs located at the project level were to devote one-third of their working hours to field visits for efficient supervision of field trials, providing timely guidance to grass root extension functionaries and also for ensuring on-the-spot solutions of the farmers problems. The findings related to information output behaviour of the extension workers and also the subject matter specialists indicated that field mobility, interpersonal interaction with the farmers, discussion with the farmers' groups, organization of field days, agro-films shows, occasional farmers' trainings etc. and use of visual aids could not be properly attended to. It was also found that VEWs neither visited the farmers' field rigidly nor they were able to see all the contact farmers in their plots. Similarly, senior SMSs visited the area only when they were informed about the problems related to their areas of specialization. Such findings are also in conformity with the findings of the case studies conducted at NIRD, Hyderabad in 1978.

The farmers processed the farm innovations with greater care prior to their actual adoption. They usually did not jump to accept the new recommendations when passed onto them. They mostly considered processing items such as economic profitability and simplicity in use of the production messages, production resources available to them, witnessing the demonstrations or plots laid by the fellow farmers and conducting trails before finally taking a decision to adopt them. As evident from the input behaviour of the farmers and also from the output behaviour to the extension-personnel, the method or result demonstrations were not properly used. Thus, the findings need to be fully realized for increased adoption rate.

The findings related to output behaviour of farmers that the knowledge levels of the contact and non-contact farmers were found to be average and below the average respectively. Similarly the farmers, in general, held unfavourable attitude towards the T and V system. The study implies that the factors responsible for creating a climate of dissatisfaction among the farmers need to be thoroughly examined and eliminated. The feedback made to extension system was also mainly confined to VEWs which were usually made unpurposively when the VEWs purposively called them during their field visits. The adoption rate in few production practices of rice i.e. concentration of high yielding varieties, timely sowing and transplanting, use of optimum seed rate, close spacing and fertilizer usage was found to be quite satisfactory. But the adoption of seed treatment, chemical weed control, plant protection chemicals and proper nursery management were found to be poor. Hence, efforts need to be concentrated to increase adoption in these production practices.

Chapter **11**

Communication strategy for Development of Poor Farmers

R. K. Pandey, Birendra Kumar, Milind C. Ahire

India lives in villages and development of rural areas in number one challenge before the nation. In order to properly comprehend the dimensions of rural development one must understand Indian villages as micro-cosm or miniature village. In India, pre-independence. A multitude of rural development programmes were launched in past includign multipurpose programmes, target area specific programmes and monopurpose rural development programmes. However, an overview of various developmental efforts revealed various missing links in their planning and implementation. Despite the series of development programmes, the much desired change among rural people did not happen in the country. Achievement has not commensurate with resources spent on them. Rural poverty is often unperceived or misperceived because of distance of the administrator and the professional from rural poor.

The essence of involving rural people in the process of their own development lies in sharing of knowledge. Sharing is not one way transfer of information, it implies an exchange between communications equals. The ultimate purpose of knowledge sharing is to empower rural people to take increasing degrees of control over their environment, over agriculture, health, habitat and other factors which so critically impinge upon the quality of life. The problem so far has been the useful sharing of knowledge has not taken place spontaneously because of cultural barriers. Development efforts have often been undetermined by incompatible communication approaches, by a clash of differing levels of educations of literacy and differing use of language. Communication for development is a response to the need to overcome such barriers.

Communication activities should reinforce the cultural identity, local values, as well as knowledge of people as an avenue to their active participation in development, communicating with rural poor is both complex and challenging because of their uniqueness in socio-economics as well as psychological characteristics. Looking to the needs of communication, it is necessary to design communication strategy for the development of rural people present study is designed with the specific objectives.

To study the socio-economic characteristics of rural farmers.

- To assess the pattern of communication for rural development programmes.
- To identify problems of farmers in rural development programmes.
- To suggest communication strategy for rural development.

Methodology :

The study was carried out in the state of U.P. the largest regions viz. eastern region and western region were selected for the study. The least developed districts from the both region namely Beharaich (Eastern region) and Badaun (Western region) were selected purposively. Further, the least developed blocks were selected from each district. Four villages were randomly selected from each block. Finally 25 farmers each were selected from eight villages to serve as respondents were interviewed with the help of protested structured interviewed schedule developed for the purpose of study. The following abbreviations of rural development programmes were used in the present paper.

Rural development programme	Abbre-viations	Rural development programme	Abbre-viations
Integrated Rural Development Programme	IRDP	Indira Aawas Yojana	IAY
Jawahar Rojgar Yojana	JRY	Rural water scheme	RWS
Ambedkar Gram Vikas	AGVY	National bio-gas development programme	NBGDP
Rural Employment Programme	RWP	Development of women programme and children in rural areas	DWCRA

Findings :

Socio-economic characteristics of rural farmers :

Designing an effective communication strategy is must to begin with knowledge of the target group. Technological messages, as well as the appropriate channel and the best format to deliver message to the greatest number of potential users should be based on the socio-economic and situation characteristics of the target audience viz. what types of communication media and formats are most appropriate for delivering the messages and for the ensuring that they will be utilized.

Table 1 revealed that majority of respondents were in the age group of 35-50 years (67%) followed by above 50 years and below 35 years respectively.

Tabble – 1. Socio-economic characteristics of poor farmers

Characteristics	Categorization	Frequency (N = 200)	Percentage
Age	Below 35 years	16	8.00
	35-50 years	134	67.00
	Above 50 years	50	25.00
Occupation-primary	Agriculture	147	73.50
	Dairy	8	4.00
	Labour	42	21.00
	Business	3	1.50
Secondary	Agriculture	46	23.00
	Dairy	14	7.00
	Labour	21	10.50
	Business	16	8.00
Caste	General	41	20.50
	OBC	81	40.50
	SC/ST	65	32.50
	Minorities	13	6.50
Land holding	Less than 1 acre	46	23.00
	1-2 acre	92	46.00
	2-3 acre	39	19.50
	More than 3 acres	23	11.50
Education	Illiterate	78	39.00
	Up to primary	86	43.00
	Up to matric	30	15.00
	Above matriculation	6	3.00

Characteristics	Categorization	Frequency (N = 200)	Percentage
Social-participation	No membership	33	16.50
	Membership in one society	134	67.00
	Membership in more than one society	33	16.50
Family type	Single family	113	56.50
	Joint family	87	43.50
Size of family	Less than 5 members	35	19.00
	5-8 members	111	55.50
	More than 8 members	51	25.50

Majority of the respondents (73.5%) having agriculture as the main occupation followed by Labour (21%), Dairy (15%), Business (1.5%) and however majority of respondents had Agriculture (23%) as subsidiary occupation followed by labour (10.5%). Business (8%) and Dairy (7%). It was also reported that majority respondents belong to backward caste (40.5%) followed by SC/ST (32.5%). General category (20.5%) and minorities (6.5%).

Majority of the respondents (46%) had 1 to 2 acres of land holding followed by below 1 acre (23%), 2 to 3 acre (19.5%) and above 3 acres (11.5%). Majority of the respondents (43.%) had education up to primary followed by illiterate (39%), up to matric (15%) and above matric (3%).

67% respondents had having membership in one society followed by no membership (16.5%) and membership in executive body (11.5%). It was also majority of the respondents (56.5%) were belonged to single family followed by joint family (43.5%). It was reported that majority of the respondents (55.5%) were having 5 to 8 members in their family, followed by more than 8 members (25.5%) and less than 5 members (19.0%).

I. Pattern of communication for rural development programme:

Awareness about rural development programme : Table 2 revealed that majority of the respondents are aware about IRDP (56%) follwed by JRY (46.5%). Thirty four had heard of AGVY followed by IAY (33%), RWS (21%), REP (20%) NBGDP (17.5%) and DWCRA (13.00%).

Table-2. Farmer's awareness and knowledge of rural development programmes

Rural development awareness programme		Knowledge		
		Goal	Target people	Benefits
IRDP	113 (56.5)	69 (34.50)	72 (36.00)	64 (32.00)
IAY	66 (33.0)	36 (18.00)	37 (18.50)	30 (15.00)
JKY	93 (46.5)	44 (22.00)	47 (23.50)	38 (19.00)
AGVY	68 (35.00)	14 (7.00)	12 (6.00)	12 (6.00)
DWCRA	26 (13.00)	8 (4.00)	9 (4.50)	4 (2.00)
NBGDP	35 (17.50)	6 (3.00)	5 (2.50)	6 (3.00)
RGP	40 (20.00)	10 (5.00)	20 (5.00)	8 (4.00)
RWS	42 (21.00)	10 (5.00)	10 (5.00)	11 (5.50)

(Figure in the parenthesis indicates percentages)

Knowledge about rural development programme - The table 2 depicts that majority of the respondents knew goals of IRDP (34.5%) followed by goals of JRY (22%), RWS (21%), IAR (18%), AGAY (7%), REP (5%), DWCRA (4%) and NBDP (3%).

As far as farmers' knowledge regarding target group of different rural development programme was concerned majority of the respondents knew about IRDP (36%) followed by JRY (23.5%), IAY (18.25%), AGAY (6%), REP (5%), RWS (5%) DWCRA (4.5%) and NBGDP (2.5%). However, incase of knowledge of benefits available in rural development programme, majority of the respondents had knowledge of IRDP (32%) follwed by JRY (19%), IAY (15%), AGAY (6%), RWS (5.5%), REP (4%), NBGDP (3%) and DWCRA (2%).

II. Sources of information first information sources among farmers:

Data pertaining to first informationsource for rural development programmes as indicated in table 3 revealed that Extension personnel (35%) were at first rank followed by Fellow farmers (23%) had second rank, Beneficiaries (22%) at third rank. Newspaper (19.5%) at forth rank, Friend/Relative (13%) at firth rank, Rural leaders (11%) had sixth rank, Radio (7.5%), Meeting (4%) and Banking (1%) and accordingly ranking had seventh, eighth and ninth rank respectively.

Table – 3. Information sources preferences, utilization of sources for detailed information, farmers perception of useful information sources and suggested information sources

Information	Information source preference as first sources	Rank	Utilization of sources for detail information	Rank	Farmers perception of useful information sources	Rank	Suggested information sources	Rank
Beneficiaries	44 (22.0)	II	62 (31)	V	32 (16.0)	V	55 (27.50)	V
Follow	46 (23.00)	IV	91 (45.50)	III	37 (17.50)	III	48 (24.00)	VI
Farmers exm. Personnel	70 (35.00)	I	116 (58.00)	I	97 (48.50)	I	47 (23.50)	VII
Newspaper	39 (19.5)	III	58 (29.00)	VI	43 (21.50)	II	32 (16.00)	XII
Radio	15 (7.50)	VII	48 (24.00)	VIII	13 (6.50)	IX	97 (48.50)	II
Friends/relatives	26 (13.00)	V	66 (33.00)	IV	27 (13.50)	VII	-	-
Meetings	8 (4.00)	VIII	23 (11.50)	IX	17 (8.50)	VIII	112 (56.00)	I
Rural	22 (1.00)	IX	99 (44.50)	II	34 (77.00)	IV	45 (22.50)	VII
Leader Bs	2 (1.00)	IX	15 (7.50)	X	32 (16.00)	VI	24 (12.00)	XIII
Posters	-	-	2 (1.00)	XII	-	-	71 (35.50)	IV
Pamphlets	-	-	4 (2.00)	XIII	4 (2.00)	XI	42 (36.00)	III
Shopkeeper	-	-	15 (7.50)	XI	9 (4.50)	X	5 (2.50)	XIV
TV	-	-	-	-	2 (1.00)	XII	41 (21.50)	IX
Munaili	-	-	-	-	-	-	73 (36.00)	III

Sources used for detailed information :

Data pertaining to table 3 revealed that Extension personnel (58%) were the most used sources, ranked at the top followed by Rural leader (49.5%) and fellow farmers (45.5%) at the second and third ranks respectively. The other sources were Friends/Relative (33%) at forth rank. Beneficiaries (31%) at fifth rank, Newspaper (29%) at sixth rank, Bank (25%) at seventh rank, Radio (24%) at eight rank and Meeting (11.5%) at ninth rank. However, sources at the bottom three places were Shopkeeper (7.5%), T.V. (3.5%) and Poster (1%) with tenth, eleventh and twelfth place respectively.

Suggested information sources :

Table 3 revealed that Meeting (56%) was the source suggested by majority of the respondents at first place followed by Radio (48.5%) and Pamphlet (36%) and second and third rank respectively. The other suggested sources in the hierarchy were Poster (35.5%) at forth rank, Beneficiaries (27.5%) at fifth rank, Fellow farmers (24%) at seventh rank, Extension personnel (23.5%) at eighth rank. Rural leaders (22.5%) at ninth rank, T.V. (20.5%) at tenth rank, Munadi (18.5%) at eleventh rank, Newspaper (16%) at twelfth rank, Bank (12%) at thirteen rank and Video (5.5%) at fourteen rank.

III. Problems in rural development :

Table-4. Major problems in rural development programme

Problems	Frequency (N = 200)	Percentage
Partial benefit	150	75
Lack of awareness about different programmes	130	65
Lack of detailed knowledge	84	42
Biasness among development functionaries	76	38
Unequal benefits among beneficiaries	72	36
Lack of discussion with general farmers	66	33
Delay in getting benefits	64	32
Complex formalities/Procedures	60	30
No training/follow-up support	48	24
Political interference in implementation	36	18

Fanners reported problems faced with regard to rural development programmes. Table 4 revealed that majority of the respondents reported Partial benefits (75%) as major problem followed by lack of awareness about rural development programmes (65%), lack of detailed knowledge (42%), biasness among development functionaries (38%), unequal benefits among beneficiaries (36%), lack of discussion with general fanners (33%), dealy in getting benefits (32%), complex formalities/procedures (30%), no training/follow up support (24%) and political interference in implementation (18%).

IV. Communication strategy :

The need - The result of the study regarding audience, their communication pattern and problems related with rural development programmes provides us a basis for formulating communication strategy for rural development. The context of the rural development calls for communication strategy to inform largest number of poor farmers about rural development programme in operation and motivate them to seek details. There is definite need to train farmers to get most advantage of these programmes.

Approach of communication :

Achievement of these objectives required focusing communication not only on the fanners but also on those other audiences who are either directly or indirectly, involved in rural development. The strategy must plan specific efforts for each of the groups. Based on behaviourial characteristics, the target audiences may be segmented into the illiterate marginal farmers and labourers on one side and literature land owning farmers of higher caste on the other. As melkote and Babbili (1986) had indicated that Action for Cultural and Political Change (ACPC) an organization of Harijan Tamil Labours in Tamilnadu initiated development of ignorant poor through a self-reliant people moved at grass root. The main vehicle of the change was the process of concientizing people about their actual situation and then actually helping them in organizing action to overcome their miseries. The crux of ACPC approach lies in the bringing group of poor together to reflect over their problems in concrete ways.

Narual (1987) had also suggested that the current development situation is not conducive for people to take initiative. They suggested development - dialogue - action - strategy to counteract low development motivation among the masses. Such a development communication strategy may focus on community involvement and self reliance through self-management of socio-economic facilities provided by the government. The strategy suggests active involvement of development functionaries and community as a whole in mutual search of solutions. It requires meeting with development functionaries and voluntary groups to discuss about discontent and expectation about on going ruraldevelopment progammes in dialogic manner so that group can be motivated for development action.

Rao (1985) suggested role of voluntary organizations in order to release creative energies of people in helping themselves. They can facilitate thinking process among villages or catalyze people into thinking and action to train people to handle their problems. They may co-operate with the government to the extent needed in order to facilitate development process. Thus, putting them together, communication for rural development requires organizational forum to bring consciousness among the disadvantaged rural groups and help them to analyze problems as well way to solve them. This may be done through a voluntary organization on the initiative of people (target people) themselves so that arrangements for sharing of experiences, discussion of problems and co-ordination with different groups and organization be possible.

The role of communication :

In communication there is no 'best' medium. No single medium by itself can meet the requirements for achieving broad coverage of fanners to inform, motivate and train them. Each medium has a particular characteristics that makes it more appropriate for one or the other of these functions. The key lies in taking advantages of the strength of each medium to complement and reinforce each others. The 'best' medium for motivating a farmer of the benefits of development programmes is, without questions another farmer who has receive the benefits and obtained demonstrable positive results.

Findings showed that most poor farmers had Radio and listened to local stations. In addition, farmers indicated that they were genuinely interested in receiving detailed information. Although, the level of illiteracy in the area was high, majority of the families had members who know to read or write. Thus, a communication strategy is proposed, one that would integrate the use of Radio to be complemented by group methods – meetings, demonstrations and traditional extension system. With this strategy, the poor farmers could receive information in their homes over Radio and seek details while periodically attending group meetings in their communities and visiting places to observe the benefits. Radio listening clubs (Rural Radio forums) have been tried in past in India. Now, with proliferating Farm and Home Units and increasing availability of economic audio-cassette recording/playing machines, development message from Radio can be recorded and played during the meetings and later discussed.

Even though, the study has indicated possibility of using certain communication source/channels and their combination for developmental awareness among rural disadvantaged. It is generally understood that details of rural development programmes are not known by the respondents. Thus, the messages should include information on goals, activities, benefits and procedure of development programmes in the simpleform so that even low literate or neo-literates can understand. Similar messages may also be conveyed through Radio, group meetings and others. However, in order to decide on format and treatment, general investigations on the prevailing forms, folk terms, songs etc. should be conducted.

References :

1. Narula, U. (1987), Priorities in development communication research in India some salient issues, Interaction vol. 5 (1) pp. 20-39.
2. Melkote, S. R. and Babbili, A. S. (1986) Socio-cultural milau of peasant in the world : A neglected factor in rural development strategies. Interaction, vol. 4 (3), pp. 1-19.
3. Rao, N. B. (1985) Proceedings of national seminar on 'communication strategies for rural development' Centre of Media Studies, New Delhi.

Chapter **12**

Information Sources Utilization Behaviour of Sericulture Enterepreneurs

S. Suriyanarayanan, G. Tamilselvi

Secriculture in India is rural based agro industry practiced mainly as a income generatin activity. Sericulture consists of both on-farm and non-farm activities and also decentralized for bringing a desired socio-economic changes at village level. It is a viable cottage industry catering to the economically small and marginal farm families, hence this enterprise is well accepted as an instrument for rural development. Entrepreneurship has now been recognized as a concept, not only vital for starting industries but also in the development of agriculture and allied activities. An entrepreneur with high information sources utilization behaviour can contact rich information sources frequently which enables him to gain knowledge about his enterprise and to take national decisions. Keeping this fact in view, the present study was taken up to study the information source utilization behaviour of sericulture entrepreneurs.

Methodology :

The present study was conducted in Dharmapuri and Pennagram blocks of Dharmapuri district in Tamil Nadu, as they had maximum population of sericulturists. From each block, three villages were selected by following the proportionate random sampling procedure. The information sources utilization behaviour was referred to as the souces/channels, where the respondent got technical information regarding his enterprise. A list of sources used by sericulture entrepreneurs was prepared in consultation with personnel of State Department of Sericulture and successfully entrepreneurs and the various sources were grouped under personal localite, personal cosmopolite and impersonal cosmopolite sources. The score for frequency of use of the sources for – seeking information was ranging from 3 to 1 for 'always', 'sometimes' and 'never' respectively. The sources for all the sources reported to have been used by the respondents for seeking information were summed up to get the total score of information sources utilization behaviour. The information sources utilized by the respondents were also ranked. The data was collected by using a well structured and pre-tested interview schedule.

Findings and discussion :

Information source utilization behaviour of sericulture entrepreneurs. The results on distribution of respondents according to their information source utilization are presented in Table 1.

Table-1. Distribution of respondents according to their information source utilization behaviour

(n = 120)

S. No.	Category	Number of respondents	Present
1.	Low	15	12.50
2.	Medium	64	53.30
3.	High	41	34.20
	Total	120	100.00

From table 1, it could be noted that more than fifty percent of the respondents (53.30 percent) had medium information sources utilization behaviour followed by 34.20 percent of the respondents with high information source utilization behaviour. Only 12.50 percent respondents had low information source utilization behaviour. The availability of information sources, formal education and better social participation of the respondents would have helped them to use the information sources better. This finding is in agreement with the findings of Ramkumar (1997).

Personal cosmopolite sources used by the respondents :

It could be seen from table 2 that 'field officers' is the most utilized sources majority of the respondents (85.83 percent) followed by training (70.00 percent), Assistant officers (78.33 percent) and field inspectors. A little more than forty percent of the respondents utilized 'Tours/field trips' for getting information. The sources viz., 'Technical Assistant', Agriculture officers (sericulture) were found to be utilized by 36.67 and 22.50 percent of the respondents respectively. The information sources found to be least utilized by the respondents were scientists (10.83%). Assistant Director of Sericulture (9.11 percent) and Joint Director of Sericulture (3.33 percent). The field officers are the grass root level workers in the state Department of Sericulture and they often visit farmers field and give technical guidance on sericulture. This condition would have enabled the sericulture entrepreneurs to contact the field officers more frequently.

Table-2. Sources of communication

S. No.	Sources/Channel	Frequency	Percent	Rank
1.	Joint director of sericulture	4	3.33	X
2.	Assistant director of sericulture	11	9.17	IX
3.	Agriculture officers (Sericulture)	27	22.50	VII
4.	Assistant agriculture officers	94	78.33	III
5.	Field inspectors	91	75.83	IV
6.	Field officers	103	85.83	I
7.	Technical assistants	44	36.67	VI
8.	Scientists	13	10.83	VIII
9.	Tours/field trips	51	42.50	V
10.	Training	84	70.00	II

Personal localite sources used by the respondents :

Table 3 reveals that the sources viz., 'relatives' and 'friends' were found to be utilized by 94.17 and 91.67 percent of the respondents respectively. The third important sources found to be utilized by 88.33 percent of the respondents were 'successful entrepreneurs' and 'input dealers'. Around sixty percent of the respondents (59.11 percent) contacted F.D.G. conveners for getting information.

Family members and village leaders were the least utilized sources by 45.00 and 27.50 percent of the respondents respectively. As the friends, relatives and neighbour are the highly personal sources and hence the respondents could contact them easily getting information. Similarly, the respondents preferred 'successful entrepreneurs' so that they can gain information on exact ways of success in their enterprise.

Impersonal cosmopolite sources utilized by the respondents :

It may be observed from the table 4 that 'Radio' was the most utilized sources by majority of the respondents (89.17 percent). 'Leaflets' and 'Newspapers' were the second and third information sources with their utilization by three-fourth (76.67 percent) and two-third (66.67 percent) of the respondents.

Table-3. Personal sources used by the respondents

(n = 20)

S. No.	Source/channel	Frequency	Percent	Rank
1.	F.D.G. conveners	71	59.17	IV
2.	Successful entrepreneurs	106	88.33	III
3.	Friends	110	91.67	II
4.	Relatives	113	94.17	I
5.	Input dealers	57	47.50	V
6.	Neighbours	106	88.33	III
7.	Family members	42	35.00	VI
8.	Village leaders	33	27.50	VII

It could be further observed that more than the proportion (55.83 percent) of respondents utilized 'magazines/other publications' for information. 'Posters/charts' and 'Exhibitions' were the next important sources utilized by 48.33 percent and 46.67 percent of the respondents respectively.

The last utilized sources were wall paintings (11.67 percent), video show/film show (20.00 percent) and Television (25.00 percent).

Radio was preferred by majority of the respondents for getting information because of is cheaper cost and accessibility. Moreover, the entrepreneurs felt that the programmes broadcasted through radio were simple to understand and full of interest. As majority of the entrepreneurs were educated, they preferred leaflets and newspapers. They felt that the published materials could be preserved and referred whenever they need information. This finding is in conformity with the findings of Govind and Perumal (2002).

Table-4. Impersonal cosmopolite used by the respondents

S. No.	Source/channel	Frequency	Percent	Rank
1.	Posters/charts	58	48.33	V
2.	Leaflets	92	76.67	II
3.	Wall painting	14	11.67	IX
4.	Radio	107	89.17	I
5.	Newspapers	80	66.67	III
6.	Magazines/other publicatiosn	67	55.83	IV
7.	Video show/film show	24	20.00	VIII
8.	Television	30	25.00	VII
9.	Exhibitons	56	46.67	VI
10.	Display boards	14	11.67	IX

Conclusions :

Among the personal cosmopolite information sources, field officers, training, assistant agril. Officer and field inspector were the most utilized sources. Majority of the respondents utilized the personal licalite sources viz. relatives, friends, neighbours and successful entrepreneurs. Radio, leaflets and newspapers were found to be most utilized impersonal cosmopolite sources by majority of the respondents. Hence, it is suggested that all these information soruces may be properly utilized for dissemination of sericulture technologies. Moreover, it is recommended that more number of printed materials especially leaflets of sericulture may be published and the coverage of sericulture programmes in both radio and newspapers should be increased so as to meet needs or sericulture entrepreneurs.

References :

1. Ramkumar, V. M. (1997). An analysis of information needs and constraints of Agricultural Entrepreneurs, unpublished, M.Sc. (Ag.), Thesis, Annamalai University, Annamalai Nagar.
2. Govind, S. and G. Perumal (2002). "Pattern of information sources utilization on IPM practices among paddy farmers", Journal of Communication Studies, 2 (20) : 62-66.

Chapter **13**

Pattern of Information Source Utilization by Paddy Farmers

Santha Govind, G. Perumal

Paddy is an important food grain in our country. The maximization of food grain production by using various chemicals and fertilizers without considering the environment gave rise to many pests and diseases complex resulting with the use of toxic pesticides. To overcome these problems, Government of India launched the Integrated Pest Management Programme (IPMP) which aimed at educating the farmers through various extension methods viz. Training, demonstration and mass media. The main objective of IPM is to empower farmers to become managers and decision makers so that they can handle and control method to maximize profits, while optimizing production. It will develop favourable attitude towards IPM, which will ultimately lead to adoption of the same. Popularization and effective dissemination of IPM play more social inputs in terms of information. Keeping this in view, a study was taken up with the objective to study the various information source utilized among IPM and Non-IPM farmers to obtain information on IPM technology for rice crop.

Methodology

The study was taken up in Tanjavur and Kumbakonam taluks of Tanjavur District as they had the maximum coverage of farmers under IPM programme. Again two villages in the selected *taluks* were chosen based on the same criteria. As per the objectives of the study, the study is to compare the information sources utilized by rice farmers belonging to IPM compact block demonstration villages with that of farmers in non-IPM respondents belonging to the same talkus and blocks that were chosen already for the selection of IPM farmers. Proportionate random sampling was adopted to select the required 100 farmers each from IPM and Non-IPM villages. Thus, a total sample of 200 respondents were studied. A list of sources used by rice farmers in the study area was prepared in consultation with extension workers, university scientists and with a few progressive farmers and the various sources were grouped under personal localite, personal cosmopolite and mass media sources. Following the procedure adopted by Ramachandran (1974) and Anthimutha (1990), information sources utilization was measured involving two dimensions, the number of sources utilized as well as technical competency. Weightage for the relevant information sources was fixed by judges opinion. The data was gathered by using a well-structured pre-tested interview schedule. The pattern of information sources utilized by IPM and Non-IPM farmers were ranked.

FINDINGS AND DISCUSSION

Pattern of personal localite source utilization

It could be seen from Table 1 that, among IPM farmers, the 'demonstration farmer' is found to be the most utilized source with its utilization by 66.00 percent while with the 'trained farmer' and 'neighbour' are found to be second and third information soucre with its utilization by 62.00 percent and 50.00 percent of IPM farmers respectively. Among non-IPM farmers, 'neighbour' is found to be the most utilized source (45.00 percent. This is followed by 'friends' and 'relatives' utilized to the extent of 40.00 percent and 38.00 percent respectively. This finding derives support from the finding of Mohankumar (1985).

Table-1 : Pattern of personal Localite source utilization by IPM and Non-IPM Farmers

Particulars	IPM farmers (n=100)			Non-IPM farmers (n=100)		
	No.	Percent	Rank	No.	Percent	Rank
Family members	25	25.00	VI	25	25.00	V
Friends	35	35.00	IV	40	40.00	II
Neighbours	50	50.00	III	45	45.00	I
Relatives	30	30.00	V	38	38.00	III
Villages leader	25	25.00	VI	28	28.00	IV
Contact farmer	20	20.00	VII	15	15.00	VI
Trained farmer	62	62.00	II	-	-	-
Demonstration farmer	66	66.00	I	-	-	-
Printed photograph	42	42.00	VI	-	-	-

Pattern of personal cosmopolite source utilization

It may be observed from the Table 2, that all the 10 personal cosmopolite sources were utilized by IPM farmers, whereas only eight information sources were made use of by Non-IPM farmers. It may be further seen while 'Assistant Agricultural Officers' is the most utilized source by nearly (95.00 percent) of IPM farmers, it is the second information source utilized by only 48,00 percent of Non-IPM farmers. This finding is line with the finding of Sisco *et. al.* (1983) and Radhakrishnan *et.al.* (1991). It is interesting to note that 'demonstration' is found to be second information source utilized by 70.00 percent of IPM farmers, whereas none of the Non-IPM farmers utilize this source. Input agencies served as the most utilized source (68.00 percent) among the Non-IPM farmers. This finding is in conformity with the findings of Amalraj and Prasad (1984). Deputy Agricultural Officers were found to be third utilized source by IPM and Non-IPM fanners respectively.

Table-2 : Pattern of personal Cosmopolite source utilization by IPM and Non-IPM Farmers

Particulars	IPM farmers (n=100)			Non-IPM farmers (n=100)		
	No.	Percent	Rank	No.	Percent	Rank
Assistant Agricultural Officers	95	95.00	I	48	48.00	II
Assistant Development Officers	58	58.00	V	24	24.00	III
Deputy Agricultural Officers	69	69.00	III	18	18.00	IV
Agricultural Officers (T&V)	38	38.00	VIII	6	6.00	V
Agricultural Officers (SMS)	35	35.00	IX	4	4.00	VII
Input Agencies	46	46.00	VI	68	68.00	I
Demonstration	70	70.00	II	-	-	-
Field trip	41	41.00	VII	-	-	-
Meeting	62	62.00	IV	5	5.00	VI
TRRI Scientists	11	11.00	X	3	3.00	VIII

Pattern of mass media source utilization

With regard to IPM and Non-IPM farmers, it may seen from the Table 3, that all the ten mass media sources were utilized by IPM farmers, whereas only 7 information sources were made use of by

Non-IPM farmers. Radio was found to be the second source of information with its utilization by 70.00 percent of IPM farmers, while it is found to be the first information with its utilization by only 50.00 percent of Non-IPM farmers. This finding is in accordance with the findings of Stuart (1988). A glance over the Table further reveals that leaflets and exhibition were the second and third information sources with its utilization by nearly three-fourth (72.00 percent) and nearly two-third (62.00 percent) of IPM farmers, while none of Non-IPM farmers utilized the same. This is in conformity with findings of Selvakumar (1988) who also reported leaflets as the most utilized source for receiving information for the control of whitefly in cotton.

Table-3 : Pattern of Mass Media Source utilization by IPM and Non-IPM Farmers

Particulars	IPM farmers (n=100)			Non-IPM farmers (n=100)		
	No.	Percent	Rank	No.	Percent	Rank
Television	22	22.00	VIII	16	16.00	III
Radio	70	70.00	II	50	50.00	I
Agricultural film	14	14.00	IX	-	-	-
Newspaper	58	58.00	IV	42	42.00	II
Leaflets	72	72.00	I	12	12.00	IV
Exhibition	62	62.00	III	8	8.00	VI
Poster/chart	48	48.00	V	10	10	V
Field board	42	42.00	VI	-	-	-
Hoarding	28	28.00	VII	-	-	-
Farm publication	14	14.00	IX	4	4.00	VII

Conclusions

There was difference in the information source utilization among IPM and Non-IPM farmers. Compact block demonstration farmers and assistant agricultural officers were found to be most utilized sources among IPM farmers. Hence, it is recommended to strengthen the knowledge on IPM among block demonstration farmers and assistant Agricultural officers, so as to help in the rapid dissemination and adoption of the technology. Further, specially designed leaflets with appropriate caption on IPM messages are suggested information materials.

REFERENCES

1. Amalraj, V. A. and M. S. Prasad 1984. "Role of Fertilizer Dealers in the Transfer of New Technology", Ind. J. Extn. Edn., 20 (3&4) : 69-70.
2. Athimuthu, P. 1990. Diagnostic study on Information Management, Learning Experience and Extent Adoption of Nutrient Use Technology for Rice Unpublished Ph.D. Thesis; TNAU, Coimbatore.
3. Mohan Kumar, V. A. 1985. An analysis of Diffusion of Selected Paddy various in Chengalpattu District. Unpublished M.Sc. (Ag.) Thesis; TNAU, Coimbatore.
4. Ramachandran, P. K. 1974. A multi variate study on Information source utilization of Big, Medium and Small Farmers. Unpublished Ph.D. Thesis; Division of Agricultural Extension, IARI, New Delhi.
5. Radhakrishanan, R. Vijayaragavan and V. Ravichandran. 1991. "Source utilization and Adoption of Grape Technology", Tamil Nadu Journal of Extension Education, 2 (4): 374-37.
6. Selvakumar, B. 1988. Information support utilization for Awareness, conviction and Adoption of cotton whitefly control measures, by contact and non-contact farmers. Unpublished M.Sc. (Ag.) Thesis; TNAU, Coimbatore.
7. Sisco, R., F. T. turpin and H. R. Potter. 1983. "Pest Management Practices of Indian Crop Farmers", Bulletin of the ESA, 29 (3) : 50-54.
8. Stuart, T. H. 1988. Participatory Communication Approaches for the Project : Integrated Pest Management Extension and Women. Paper Presented at the Conference on Appropriate Technology for Rural Women, New Delhi, India.

Chapter 14

Predictive Values of the Selected Factors in Contributing Towards the Variation in the Effectiveness of Communication

B.N. Choudhary and C. Prasad

Fertilizer is one of the critical inputs, on which, production of food grains depends considerably. Although, efforts are being made continuously to educate farmers for use of balance fertilizers, yet we are far off from the reach of the desired targets.

With the growth of private, cooperative and Govt. sector fertilizer plants the educational scene has witnessed significant change in approach of Reaching Out. The article examines the significant elements of communication in relation to this National endeavour. It highlights the points such as: trustworthiness of communicator, extent of use of channels, communication skills of Communicator and feedback process etc. which influence changes in the decision making process, at the level of farming community.

To talk about effective communication is to talk about how people learn, how much knowledge they have gained, how much their attitudes have changed, how much skills they have acquired, and what action they have decided to take due to changes in their knowledge, attitudes and skills. Communication of highly technical information needs to be seen not only as a simple process occurring between the communicator and the communicate, rather a highly sophisticated phenomenon involving many factors which in all permutations and combinations create a very complex system of forces accelerating or retarding the effectiveness of communication.

The success of communication act had mostly been studied in respect of the adoption of agricultural innovations. Onlya few studies have been made to find out the factors influencing the effectiveness of communication in adoption of agricultural innovations. In the present study, the effectiveness of communication has been studied in respect of fertiliser promotion, a basic input for agricultural production.

The following two objectives were set for the study :

i) to identify and select some important factors influencing the effectiveness of communication in fertiliser promotion; and

ii) to estimate the predictive values of the selected factors in contributing towards the variation in the effectiveness of communication in fertiliser promotion.

Assumptions

A number of assumptions have been based on theoretical framework of the study and review of literature. These assumptions are :

- Communication is the process having five elements: source, message, channel, receiver, and feedback and in absence of any one element, the process will remain incomplete;
- Communication to promote three types of fertilisers, viz., nitrogenous, phosphatic and potassic and their uses have taken place in the area of the study;
- There are factors in each element of the communication process which influence the effectiveness of communication in fertiliser promotion and these factors are identifiable;
- The perception about the relative importance of identified factors of the persons engaged in fertiliser promotion, in general, would be valid and some important factors can be selected based on the relative importance of these factors which are associated with the effectiveness of communication in fertiliser promotion;
- The effectiveness of communication can be measured in respect of changes in the farmers' behavioural components, viz., knowledge, attitude, skills and symbolic adoption;
- A composite measure of effectiveness of communication in fertiliser promotion is possible by adding the scores on all the four components of behaviour per unit of communication sensitivity;
- The variation in communication sensitivity from one individual to another is possible; and

- The factors in each element of communication related to the effectiveness of communication in fertiliser promotion can be objectively measured and quantified and their relationship and contribution towards explaining the variation in the effectiveness of communication can be assessed.

Conceptual Model

Based on the review of past studies, judges ratings of the importance of factors and the assumptions made in this study, a conceptual model was developed and the same is depicted in Fig. 1. A brief description of the relationship between various elements of the model which was conceptualised in the present study are given below :

- In the conceptual model, the communication process has five distinct elements, viz., source or communicator, message, channel, receiver and feedback. Sources are the person or a group of persons who communicate the information of fertilisers and their application. The sources are: Friends, Neighbours, Village Leaders, Progressive Farmers, Fertiliser Depot Managers, Village Fertiliser Salesmen, Block Agricultural Officers, Project Executive Officers, Village level workers and Agricultural Scientists. Message is the information related to fertilisers and their uses that the sources wish their audience or uses that the sources wish their audience or receiver to receive, understand, accept and act upon.
- Channel is the bridge between the source and receiver of the messages. The channels in the present study are Demonstration, Radio, Film show, Puppet show, Exhibition, Fertiliser Festivals, Field Days, Village Meetings, Folders and Posters. Receivers are the persons who receive the message from the source. The farmers are the receivers of the message. The feedback is the response of receiver on the message communicated by source.
- The effectiveness of communication in fertiliser promotion, measured in terms of changes in the farmer's behavioural elements, that is, knowledge, attitude, skill and symbolic adoption, are influenced by a number of factors in each element of the communication process. These factors are; communication skill of communicator, trustworthy and expertness of communicator, Frequency of contact of communication, attitude of communicator towards receiver, attitude of communicator

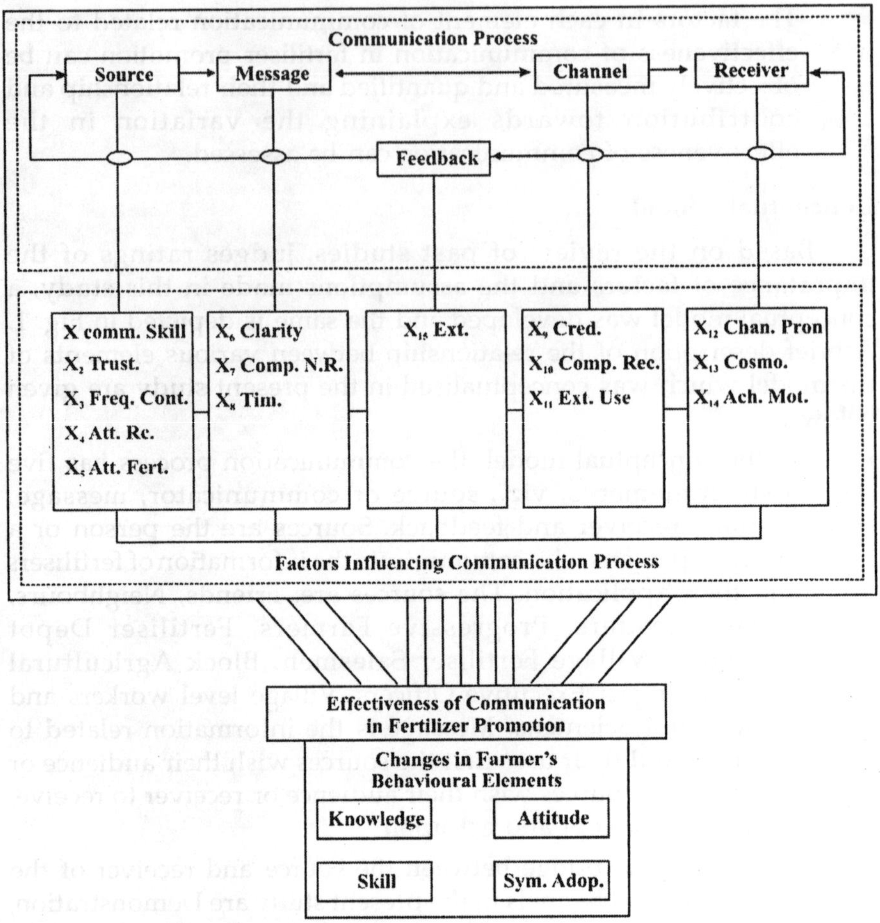

LLGEND

X_1 Comm. Skill = Communication Skill of
X_2 Trust. = Trust worthiness of
X_3 Freq. Cont. = Frequency of Contact of
X4 Att. Rec. = Attitude Towards Receivers
X_5 Att. Fert. = Attitude Towards Fertilizers
X_6 Clarity = Clarity of
X_7 Comp. Nr. = Compatibility with the need of receiver
X_8 Tim. = Timeliness of
X_9 Cred. = Credibility of
X_{10} Comp. Rec. = Compatibility with Receiver
X_{11} = Ext. Use = Extent of Use of
X_{12} Chan. Pron. = Change Proneness of
X_{13} Cosmo = Cosmopolitanisms of
X_{14} Ach. Mot. = Achievement of Motivation of
X_{15} Ext. = Extent of Feedback

Figure 1. Conceptual Model of Effectiveness of Communication in Fertilizer Promotion

towards fertilisers, clarity of message, compatibility of message with the need of receiver, timeliness of message, credibility of channel, compatibility of channel with receiver, extent use of channels, change-proneness of receiver, cosmopoliteness of receiver, achievement motivation of receiver and extent of feedback.

Further, all the scores of farmers on these four behavioural elements are clubbed to give Fertiliser Promotion Index. The same index was calculated in terms of per unit of communication sensitivity of individual in order to get effectiveness of communication in fertiliser promotion.

Research Method

The study was conducted in the Purnea and Saharsa Districts of Kosi Division in Bihar. All the blocks of the Purnea and Saharsa districts were separately classified into two categories on the basis of their adoption indices in respect to fertiliser (NPK) consumption per hectare of arable land. One block from each of the two categories, viz., low and high, was randomly selected from each of these two districts. Thus, finally four blocks, - namely, Krityanand Nagar and Raniganj in Purnea and Madhepura and Singhesh-warsthan in Saharsa district, were selected.

All the villages of each of these four selected blocks were classified again into two categories on the basis of their adoption indices in respect to total fertiliser consumption (NPK) per hectare of cultivated land. Two villages, one each from low and high fertiliser consumption categories from each one of the four selected blocks were randomly selected. These villages were : Gangeli and Islampur in Krityanand Nagar, Pachira and Kalabalua in Raniganj; Bhiraki and Tuniahi in Madhepura; and Saropatti and Sabaila in Singheswarsthan.

Total farmers of each of these eight selected villages were classified into four groups, viz., large, medium, small and very small on the basis of size of their holdings. A proportionate random sample was drawn from each group in such a way that the total sample respondents (farmers) do not exceed 25 in any village. In this way, 200 farmers were selected which formed the sample of this study.

Fifteen independent variables, namely, communication skill of communicator, trustworthiness of communicator, frequency of contact of communicator, attitude of communicator towards receiver, attitude of communicator towards fertilisers, clarity of message, compatibility of message with the need of receiver, timeliness of message, credibility of channel, compatibility of channel, extent of use of channel, change proneness of receiver, cosmopoliteness of receiver, achievement motivation of receiver and extent of feedback were selected out of 70 independent variables, on the basis of their importance as rated by the extension personnel engaged in fertiliser promotion.

The empirical measures were developed to quantify all the variables, both dependent and independent under study. The dependent variable, effectiveness of communication in fertiliser promotion, was measured by developing an index which included knowledge about fertilisers and their application, attitude towards fertilisers, skill in fertilisers application and symbolic adoption for each unit of the communication sensitivity of the farmers. Knowledge of farmers about fertilisers was measured by knowledge test developed by Choudhary and Prasad (1977) and attitude of farmers towards fertiliser use was measured by Attitude Test developed by Choudhary and Prasad (1978). For measuring skills in fertiliser application and symbolic adoption, a skill test and an index of symbolic adoption were developed in the present study. The communication sensitivity of farmer was measured with the help of Jha's (1968) communication sensitivity test. The Fertiliser Promotion was measured by an index developed by Choudhary and Prasad (1979).

The communication skill of communicator was measured with the help of Singh's (1965) rating scale. Haque's (1971) rating scale was used to measure clarity of message. A rating scale developed by Jha (1968) was used to measure compatibility of message with the need of receiver. For measuring change-proneness of receiver, Moulik's (1965) self-rating scale was used. Ernest's (1973) rating scale was used for measuring cosmopoliteness of receiver. The achievement motivation of receiver was measured with the help of projective test developed by Pal and Singh (1973). For other independent variables, namely, trustworthiness of communicator, degree of contact of communicator with receiver, attitude of communicator towards receiver, attitude of communicator towards fertilisers, timeliness of message, credibility of channel, compatibilityof channel with receiver and extent use of channels, a schedule was specially developed in the present study using rating technique. All scales were pre-tested in the area and their reliability and validity were ascertained before use.

Data were collected by interviewing the farmers personally. The results were analysed and interpreted by using zero-order correlation, multiple correlation and regression, t-test, standard partial regression co-efficients, and analysis of variance techniques.

Findings

To examine the individual relationships between the dependent variable (Y) and the independent variables (X_1 .... X_{15}), correlation coefficients (r - values) were computed. The correlation coefficients, (r - values) thus worked out were tested for their statistical significance with the help of the table value of r for 198 degrees of freedom and at 0.05 level of significance and they are presented in Table 1.

Table-1. Relation of dependent variable (Y) with the independent variables (X_1 ..... X_{15})

Independent variable	Coefficients of correlation (r)
X_1 =Comm. Skill	0.6498**
X_2 = Trust.	0.8989**
X_3 = Freq. Cont.	0.3789**
X_4 = Att. Rec.	0.8109**
X_5 = Att. Fert.	0.8213**
X_6 = Clarity	0.7939**
X_7 = Comp. N.R.	0.6145**
X_8 = Tim.	0.8318**
X_9 = Cred.	0.3687**
X_{10} = Comp.	0.5337**
X_{11} = Ext. Use	0.8009**
X_{12} = Chan. Pron.	0.5045**
X_{13} = Cosmo.	0.8340**
X_{14} = Ach. Mot.	0.6872**
X_{15} = Ext.	0.6542**

**Significant at 0.1 level of probability.

Critical value of 'r' for 198 degrees of freedom and at 0.05 level = 0.139; and At 0.01 level = 0.182

It may be observed from Table 1 that the individual correlation coefficients (r) between the farmers' scores on effectiveness of communication in fertiliser promotion (Y) and communication skill of the communicator (X_1), trustworthiness of communicator (X_2), frequency of contact of communicator (X_3), attitude of communicator towards receiver (X_4), attitude of communicator towards fertilisers (X_5), clarity of message (X_6), compatibility of message with the need of the receiver (X_7), time-lines of message (X_8) , credibility of channel (X_9), compatibility of channel with receiver (X_{10}) extent of use of channels (X_{11}), change-proneness of receiver (X_{12}), cosmopoliteness of receiver (X_{13}), achievement motivation of receiver (X_{14}) and extent of feedback (X_{15}) and positive and significant at 0.01 level of probability.

Table-2. Coefficient of partial regression and multiple correlation with fifteen independent variables (X_1 .... X_{15})

Independent variables	Coefficients of partial regression (bi)	Calculate values of "t"	R^2
X_1 (Comm. Skill)	0.6406	4.4701**	
X_2 (Trust.)	1.1324	5.3017**	
X_3 (Freq. Cont.)	0.1887	2.5641*	
X_4 (Att. Rec.)	0.4882	1.1015	
X_5 (Att. Fert.)	0.4434	1.2634	
X_6 (Clarity)	-0.2807	-0.1289	
X_7 (Comp. NR)	-0.0339	-0.1833	0.9177
X_8 (Tim)	0.4049	0.8991	
X_9 (Cred.)	0.5025	1.4995	
X_{10} (Comp.)	0.5025	1.4995	
X_{11} (Ext. Use)	1.0357	4.1970**	
X_{12} (Chan. Pron.)	0.9139	2.8470**	
X_{13} (Cosmo.)	-0.1194	-0.2985	
X_{14} (Ach. Mot.)	0.8713	5.4243**	
X_{15} (Ext.)	0.3864	2.2415*	

Critical value of t at 0.05 level = 1.9600, and

At 0.01 level = 2.5758.

*Significant at 0.05 level

** Significant at 0.01 level

The coefficient of partial regression of all the 15 independent variables (X_1 ..... X_{15}), their corresponding r-values and square of multiple correlation coefficient (R^2) were calculated on computer and the same are presented in Table 2. These were statistically tested at 0.05 level for their significance.

It may be observed from Table 2 that the partial regression coefficients of only several independent variable, viz., communication skill of communicator (X_1), trustworthiness of communicator (X_2), frequency of contact of communicator (X_3), extent of use of channels (X_{11}) change-proneness of receiver (X_{14}); and extent of feedback were significant.

The values of square of multiple correlation (R^2) being 0.9177, it can be inferred that the fifteen independent variables jointly contributed towards about 92 per cent of the variation in the effectiveness of communication in fertiliser promotion.

The coefficient of multiple correlation (R^2) was also tested for its statistical significance with the help of analysis of variance test as shown in Table 3.

Table-3. Significance of the coefficient of multiple correlation with fifteen independent variables (X_1.....X_{15})

Source	d.f.	s.s.	M.S.S.	F
Regression	15	65872.00	4583.46	137.66**
Residual	184	6162.28	33.29	
Total	199	74878.28	4616.75	

** Significant at 0.01 level of significance.

Critical value of F for 15 and 184 degrees of freedom and at 0.05 level = 1.7174, and at 0.01 level = 2.1396.

The F value as shown in Table 3 is the ratio of regression mean sum of squares to the residual mean sum of squares which was 137.66 for 15 and 184 degrees of freedom. It was significant at 0.01 level of probability which indicated highly significant association of fifteen independent variables with the dependent one, that is, effectiveness of communication in fertilizer promotion, when all the independent variables were functioning jointly.

The partial regression coefficients of attitude of communicator towards receiver (X_4); attitude of communicator towards fertilisers (X_5); clarity of message (X_6); compatibility of message with the need of receiver (X_7); timeliness of message (X_8); Credibility of channel (X_9); Compatibility of channel (X_{10}); and cosmopoliteness of receiver (X_{13}) were found to be statistically non-significant (Table 2), The multiple correlation coefficients (R) and coefficients of partial regression(bi) were, therefore, calculated for only significant independent variables (X_1, X_2, X_3, X_{11}, X_{12}, X_{14} and X_{15}) and the same are presented in Table 4.

Table – 4. Coefficient of partial regression and multiple correlation with seven independent variables (X_1, X_2, X_3, X_{11}, X_{12}, X_{14}, X_{15})

Independent variables	Coefficient of partial regression	Calculated value of 't'	R^2
X_1 (Comm. Skill)	0.7138	5.0540**	
X_2 (Trust.)	1.4309	7.2413**	
X_3 (Freq. Cont.)	0.2335	3.1763**	
X_{11} (Ext. Use)	1.1662	6.6648**	0.9063
X_{12} (Chan. Pron.)	1.3857	4.5727**	
X_{14} (Ach. Mot.)	1.0518	7.8114**	
X_{15} (Ext.)	0.5139	2.9809**	

Critical value of t at 0.05 level - 1.9600 and at 0.01 level = 2.5758

** Significant at 0.01 level of probability

It is apparent from Table 4 that the coefficients of partial regression of communication skill of communicator (X1); trustworthiness of communicator (X2); frequency of contact of communicator (X3); extent of use of channels (X11); change-proneness of receiver (X12); achievement motivation of receiver (X14) and extent of feedback (X15) were significant at 0.01 level of probability. These seven factors were associated with the dependent variable, effectiveness of communication in fertiliser promotion, to the extent they jointly contributed for 90.63 per cent variation in the effectiveness.

The statistical significance of the square of coefficient of multiple correlation (R^2) of seven independent variables (X_1, X_2, X_3, X_{11}, X_{12}, X_{14}, and X_{15}) on effectiveness of communication in fertiliser promotion (V) was further tested by means of the analysis of variance and presented in Table 5.

Table – 5 Significance of the coefficient of multiple correlation with seven independent variables (X_1, X_2, X_3, X_{11}, X_{14} and X_{15})

Source	d.f.	s.s.	M.S.S.	F
Regression	7	67815.00	9687.85	263.34**
Residual	192	7063.28	36.78	
Total	199	74978.28		

Critical value of F for 7 and 192 degrees of freedom and at 0.05 level = 2.0532; and at 0.01 level = 2.7348

**Significant at 0.01 level of probability.

The F value in Table 5 indicates that the square of coefficient of multiple correlation ($R^2 = 0.9063$) of the seven independent variables, $X_1, X_2, X_3, X_{11}, X_{14}$ and X_{15} on dependent variable (r), that is, effectiveness of communication in fertiliser promotion, was highly significant at 0.01 level of probability. A comparative study of Tables 2 and 4 revealed that after elimination of eight variables ($X_4, X_5, X_7, X_8, X_9, X_{19}$ and X_{13}) from the second regression analysis, there was virtually no difference in the value of multiple correlation coefficient. The square of multiple correlation coefficient with 15 independent variables was 0.9177, whereas, it was 0.9063 with seven variables. It may, therefore, be concluded that the eight variables having non-significant partial regression coefficients, in Table 2 did not make much contribution towards the variation in effectiveness of communication in fertiliser promotion.

Further analysis involved the testing of the significance of loss in sum of squares and finally in predictability by dropping these seven ($X_1, X_2, X_3, X_{11}, X_{12}, X_{14}$ and X_{15}) variables one by one. Significance of loss was tested with the help of analysis of variance technique and for each of the seven factors, F - values were computed and the same were presented in Table 6.

From Table 6 it may be noted that corresponding F - values by dropping $X_1, X_2, X_3, X_{11}, X_{12}$, X14 and X_{15} were 25.54, 52.43, 10.09, M.S1 , 20.90, 61.01, and 8.88 which were statistically significant for 1 and 192 degrees of freedom and at 0.01 level of probability. Hence, it can be concluded that each one of these seven variables individually contributed towards the variation in the effectiveness of communication in fertiliser promotion(Y). These seven independent variables were:

X_1 = Communication skill of communicatory;

X_2 = Trustworthiness of communicator;

X_3 = frequency of contact of communicator;

X_{11} = extent of use of channel;

X_{12} = change-proneness of receiver;

X_{14} = achievement motivation of receiver; and

X_{15} = extent of feedback.

After getting the significant factors which contributed towards the variation in the effectiveness of communication in fertilizer promotion, it was felt necessary to find out the relative importance of these seven factors in explaining the variation in, the dependent variable (Y). The relative importance of each of the seven factors was established by calculating standard partial regression coefficients of each of the factors and they were presented in Table 7.

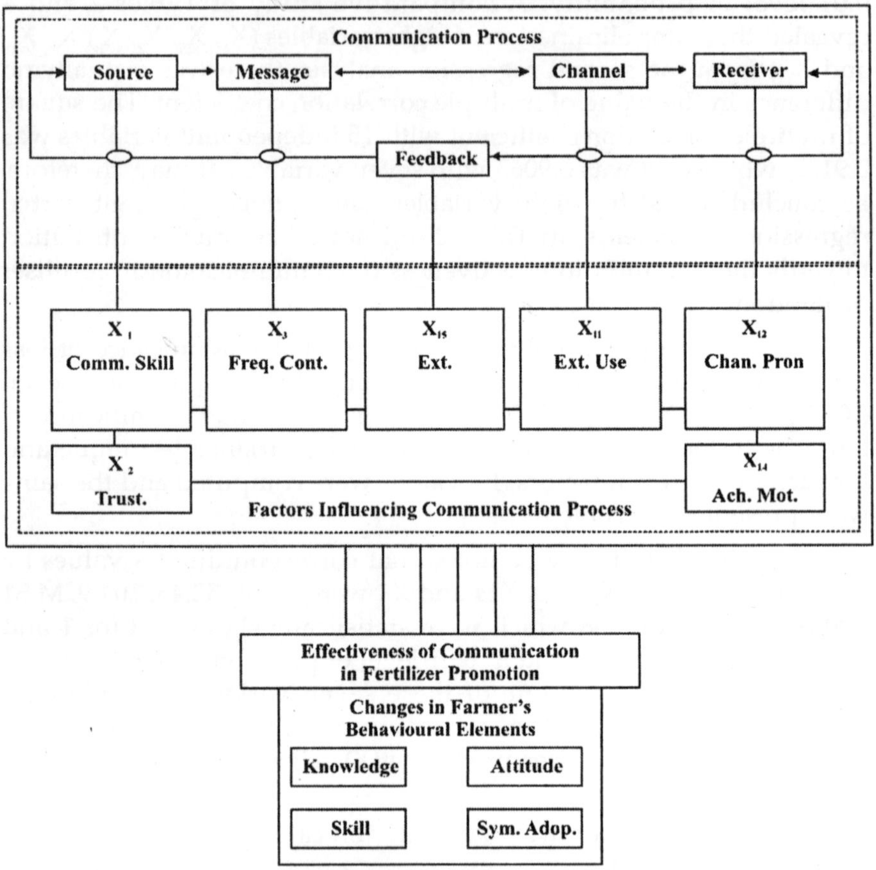

LLGEND

X_1 Comm. Skill = Communication Skill of
X_2 Trust. = Trust Worthiness of
X_3 Freq. Cont. = Frequency of Contact
X_{11} Ext. Use = Extent of use of
X_{12} Chan. Pron = Change Proneness
X_{14} Ach. Mot. = Achievement Motivation
X_{15} Ext. = Extent of Feedback
Sym. Adop. = Symbolic Adoption

Figure – 2. Model of Effectiveness of Communication in Fertilizer Promotion on the Basis of Findings.

Table-7. Standard partial regression coefficients of the seven significant independent variables (X_1, X_2, X_3, X_{11}, X_{14}, and X_{15})

Independent variables	Coefficients of partial regression	Standard partial regression coefficients	Rank order
X_1 (Com. Skill)	0.7138	0.1458	IV
X_2 (Trust.)	1.4309	0.3373	I
X_3 (Freq. Cont.)	0.2335	0.0822	VII
X_{11} (Ext. Use)	1.662	0.2424	II
X_{12} (Chan. Pron.)	1.3857	0.1168	V
X_{14} (Ach. Mot.)	1.0518	0.2393	III
X_{15} (Ext.)	0.5139	0.0890	VI

The "beta weight" or standard partial regression coefficients of the seven factors indicated the factors hierarchy as under :

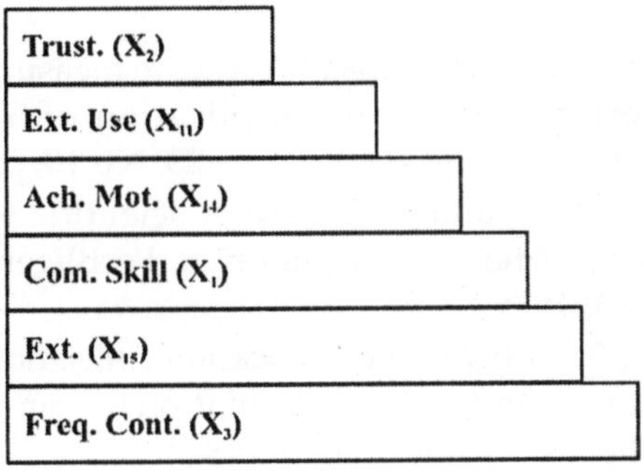

Implications

The present investigation was purposely undertakenwith a view to give a fillip to the on-going fertiliser promotion programmes. On the basis of the field observations and earlier studies it was assumed that there were several lacuna in the fertiliser adoption process and understanding of the same which was imperative on the part of the fertiliser promotors and administrators.

It was interesting to find that almost 91 percent of the variation in the effectiveness of the communication in fertiliser promotion was explained by 15 independent variables studied in this investigation. And out of these fifteen factors, only seven independent variables, namely, trustworthiness of communicator, extent of use of channels, achievement motivation of receiver, communication skill of communicator, change-proneness of receiver, extent of feedback and frequency of contact of communicator in order of their importance, explained 90 per cent of the variation in the effectiveness of communication in fertiliser promotion. This clearly implies that the extension workers engaged in the fertiliser promotion programme must pay considerable attention to these variables and utilise them suitably to popularise and promote the fertiliser promotion programme.

These findings will be helpful in developing a sound fertiliser promotion strategy in Kosi Division of Bihar. This also opens the eyes of fertiliser communicators, for they should not be complacent in communicating fertilizer information which is very basic for agricultural production.

References

1. Choudhaiy, B. N. and C. Prasad A test to measure knowledge in relation to fertilizer use, Fertilizer Marketing News, June Issue No.8, pp 1-5, 1977.
2. Chaudhaty, B.N. and C. Prasad A scale to measure attitude of farmers towards fertilizer use - An application of Scalegram Analysis Technique, fertilizer News, Volume 23, No.10, pp 1-15, 1978.
3. Choudhary, B. N. and C. Prasad A scientific tool for an objective measurement of fertilizer promotion, Fertilizer News, Volume 24, No.10, pp 1-4, 1979.
4. Ernest, R. S., A study of Communication Utilisation Behaviour of Small and Big Farmers and Its Implementation to communication strategy. Ph.D. Thesis, I.A.R.I., New Delhi., 1973.
5. Haque, S.M.S., A study on Communication of Agricultural information and the Extent of Distortion Occurring from District to Village Level Workers in Selected I.A.D.P. Districts. Ph.D. Thesis. The University of Udaipur, Udaipur, Rajasthan, 1971.
6. Jha, P.M., A critical Analysis of F actors Associated with Communication Fidelity with Reference to High Yielding Programme. Ph.D. Thesis. I.A.R.I., New Delhi, 1968.
7. Mouiik, T. K., A study of the Predictive Values of Some Factors of Adoption of Nitrogenous Fertilizers and the Influence of Sources of Information on Adoption Behaviour. Ph.D. Thesis, I.A.R.I., New Delhi, 1965.
8. Pal, Kumltum and K. N. Singh, Measurement of Achievement Motivation of Delhi Farmers by T.A.T., Paper (unpublished), 1973.
9. Singh, Y. P., A Study of Communication Networks in Sequential Adoption and Key-communicators. Ph.D. Thesis, IARI, New Delhi, 1965.

Chapter **15**

Information Role Performance of Farm Women in Different Farming System

G. Tamilselvi, M. Vetriselvan

Several studies indicate that women have been playing important role in making decisions. According to Sundarajan (1985), 60 percent of farm decisions were influenced by women. Gurusamy (1987) reported that more than 50 perecent of farm women, who are involved in rational decision making process of new practices, adopt the practices of such process and pave way for increasing production in agriculture. Information on new technologies is the basis for farm women to take rational decisions. Hence, it was felt necessary to study about the information roles performed by farm women in varied farming systems viz., wetland and dry land farming systems because the farm women of these two systems differ in their socio-personal and economic aspects.

Methodology :

This study was conducted in two different farming system viz., wetland farming system of Trinelveli district and dry land farming system of Tuticorin district of Tamil Nadu. Among the sample size of 120, sixty were selected from wetland farming system and 60 from dry land farming system. The respondents were selected by using proportionate random sampling technique. The information roles include information collection, information storage and information communication areas.

Twelve information roles were identified and referred to 30 judges for assigning weightage ranging from 0-10 according to their importance for decision making. The mean weightage was taken as the 'role weightage'. Three performance categories viz., self-doing, assisting and reminding were identified for studying the information role performance. An arbitrary scaring of 3, 2, and 1 were assigned to self-doing, assisting and reminding respectively. Each respondent was asked to indicate the roles performed by her. She was also requested to indicate the performance category under which each role was performed.

Measuring the actual information role performance of farm women:

The product of the role weightage and the performance category weightage formed the individual information role performance score. This was obtained individually for all the information roles reported to have been performed by each respondents. Such score when added for all the roles resulted in the 'Actual information role performance score'.

Computing the maximum information role performance :

The sum total of the role weightages for all the 12 roles multiplied by 3, the maximum performance category weightage resulted in the 'maximum information role performance score'.

The actual information role performance score was divided by the 'maximum information role performance score' which resulted in 'information role performance score' of the respondents.

Mean ± 1.96 SE, independent 't' test and zero order correlation were the statistical tools used for the study.

Result and discussion :

Information roles selected with their weight less :

The information roles selected for the study and their weightlessness are given in table 1.

A perusal of table 1 reveals that a maximum weightage of 6.2 was given to the role 'viewing agricultural news in T.V. followed by participating in discussion groups (5.9) and listening agricultural news in radio. Television is the media which has both audio and visual effect and thereby enables more retention of message, which in turn may enable the farm women to take rational decisions. Similarly, radio provides regular agricultural programmes thereby acts as a more accessible media for getting information. Discussion is an opportunity, which enable the farm women to share their ideas, knowledge and feeling and hence it might be assigned with more weightage.

S. No.	Roles	Weightage
1.	Listening agricultural news in radio	5.8
2.	Viewing agricultural news in T.V.	6.2
3.	Participating in discussion groups	5.9
4.	Participating in village meetings	4.1
5.	Reading newspaper	4.9
6.	Reading from journals, folders and leaflets	3.6
7.	Participating in training programmes	4.9
8.	Listening video cassettes	4.2
9.	Viewing agricultural films and slide-shows	3.4
10.	Classifying information based on profession and technology	3.8
11.	Preserving the information	4.6
12.	Giving information whenever there is need	4.7

Categorization of farm women on their information role performance

The frequency distribution of farm women of wetland and dry land farming systems on their information roleperformance are given below :

Table-2. Frequency distribution of farm women on their information role performance

S. No.	Category	Wetland system		Dry land farming system	
		Number (n = 60)	Percent	Number (n = 60)	Percent
1.	Low	38	63.33	49	81.67
2.	Medium	8	13.33	-	-
3.	High	14	23.33	11	18.33
	Total	60	100.00	60	100.00
	Mean	24.10		11.27	

Difference between the means 12.38

't' value 3.298**

** - significant at 0.01 level of probability

It is interesting to see from table 2 that majority of both the categories of respondents were with low information role performance, the frequency distribution of dry land farm women was more (81.67%) than in wetland farm women (63.33%). When no one was seen under medium performance category in the case of dry land farm women, it was 13.33 percent for wetland farm women. The frequency of respondents under high performance category was 23.13 percent for dry land farming system. It may be seen from the same table that the mean score of wetland farm women was more than that of dry land farm women. The difference between means was also highly significant.

Based on the mean scores, it may be stated that the wetland farm women were high in the information role performance than dry land farm women. From this, it may be concluded that the information role performance of farm women differed significantly in different farming system.

It is obvious that the knowledge of highly wetland farm women who had greater extension contact, more mass media participation and high information seeking behaviour, would have contacted the credible information sources more frequently and hence performed more information roles in a better way than the dry land farm women, who would not have had such opportunities because of their lack of knowledge and extension contact.

Relationship of characteristics of farm women with their information role performance :

It is interesting to see from table 3 that of the 14 characteristics studied, 7 of wetland farm women and 9 of dry land farm women showed positive and significant relationship with their role performance. Of all significant correlation values, farm size of wetland farm women and social participation and perception on information management of dry land farm women were significant at 5 percent level probability. The other correlation values of both the groups of farm women were non-significant.

From this, it may be concluded that better the education, bigger the farm size, higher the extension contact, greater the mass media participation, more the risk orientation, higher the information seeking behaviour and greater the knowledge about cultivation practices of wet land farm women, the higher would be their information role performance irrespective of their age, caste, family status, farming experience, social participation, socio-economic status and perception on information management. In the case of dry land farm women, better the education, bigger the farm size more the social participation, higher the extension contact, larger the mass media participation, more the risk orientation, greater the information seeking behaviour and

Table-3. Correlation of characteristics of wetland and dry land farm women with their information role performance

S. No.	Variables	'r' value	
		Wetland farming system (n = 60)	Dry land farming system (n = 60)
1.	Age	-0.893 NS	0.1447 NS
2.	Education	0.3496**	06597**
3.	Caste	-0.0143 NS	0.1431 NS
4.	Family status	0.1414 NS	-0.0041 NS
5.	Farm size	0.2831*	0.4209**
6.	Farming experience	-0.2146 NS	0.1255 NS
7.	Social participation	0.2239 NS	0.2927*
8.	Socio-economic status	0.2304 NS	0.1023 NS
9.	Extension contact	0.7535**	0.5288**
10.	Mass media participation	0.6564**	0.5686**
11.	Risk orientation	0.3777**	0.4553**
12.	Information seeking behaviour	0.8177**	0.3545
13.	Knowledge about cultivation practices	0.7135**	0.3774**
14.	Perception on information management	0.2148 NS	0.2675*

**Significant at 0.01 level of probability
*Significant at 0.05 level of probability
NS non-significant

more the perception on information management, higher would be their information role performance irrespective of their age, caste, family status, experience and socio-economic status.

As already operationalised elsewhere, the information roles includes information collection, information storage and information communication areas. Hence, the information would be the basis for performance of role.

Higher education of farm women of both the categories would have paved way for getting them exposed to various sources of information, which in turn would have enabled them to perform the information roles. This may be the possible reason for the reported positive and significant relationship of their education and their information role performance.

Farm size showed a positive and significant relationship with information role performance among both groups of farm women. Big farmers might have had ample chances to know about more information. Such a condition would have helped them to perform more of information roles. This could be seen from the positive and significant inter correlation value of farm size of wetland farm women with social participation and that of dry land farm women with social participation, extension contact, mass media participation, information seeking behaviour and knowledge about cultivation practices.

The positive influence of extension contact on information role performance of both the categories of farm women needs no explanation. Such a condition would have enabled them to perform information role in a better way.

Mass media participation otherwise means the respondents hearing radio, viewing T.V. reading/hearing the newspapers, magazines, bulletins, visiting exhibitions, attending meeting and engaging in post participation activities. Such a condition would have provided them more chance to gain more information better. Hence, mass media participation of both categories of farm women would have shown positive and significant relationship with their information performance as seen.

Risk orientation showed highly significant relationship with information role performance among both the groups of farm women. More risk orientation otherwise means their readiness to accept the new ideas. Such a state of mind would have naturally prompted them to seek for more information from different sources and hence would have enabled them to effectively perform the information roles.

Positive and significant relationship of information seeking behaviour and knowledge about cultivation practices of both the groups of farm women with their information role performance in this study is quite understandable. Information seeking behaviour would have exposed them to rich information sources culminating in the accumulation of understood information in the form of knowledge. This reveals a favourable state of condition for effective information role performance and hence would have resulted in the positive and significant relationship as reported.

Positive and significant relationship between social participation and information role performance of dry land farm women may be due to the interaction chances provided by social participation for sharing of information which would have helped in the information role performance. But in the wetland farm women the homogeneity in social participation S.D. value of 3.71 might be the probable reason for the reported non-significant relationship between social participation and their information role performance.

Perception on information management of dry land farm women showed a positive and significant relationship with information role performance. More perception on information management otherwise means the receipt of information in a precise, clear, complete and understandable manner. It would have enabled their information role performance. The wetland farm women were homogenous (S.D. 2.98) in their perception on information management. Hence, it would have shown non-relationship. The similar trend in the relationship of perception on information management with information role performance could be seen even among wetland farm women as shown by high correlation value, but however not upto significant level. This may be due to the fact that the information role performance might be mostly due to their contact with extension agency and information seeking behaviour among wetland farm women as revealed by very high correlation value of these characteristics in table 3 rather than by their perception on information management.

Conclusions :

The wetland farm women were high in the formation role performance than dry land farm women. Lack of information sources like radio, T.V. and lack of knowledge may be the reasons for poor performance of dry land farm women. This needs special alteration. To ensure better impact of community, media sets on farm women farmer discussion groups, women tele clubs and women farm science centers are suggested.

Chapter **16**

Communication of Agricultural Information within A Village Social System: A Model of Farm Information Flow

O. K. Gogoi and Raghbir Singh

There are a number of communication models viz. hypodermic needle model, one-step flow model, two-step flow model and multi-step flow model which help to conceptualize and to understand the phenomena related to mass communication. Each of the models has strong as well as weak points and can best be applied to specific mass communication situations. The multi-stepflow model incorporates all other models and mass communication researchers give much credence to it. But, it has the shortcoming of being least specific i.e. the elements and/or steps of the communication process are not specified in the model. In an attempt to meet the gap and especially applicable to diffusion of farm informationwithin a village, a model has been proposed. It will help to conceptualize and to get a comprehensive mental view of the farm information flow process within a village social system. It is modified version of the two-step flow model (Lazarsfeld and others, 1944) and few of the concepts used in diffusion researches of farm information have been incorporated in it.

The answers to the questions

i) How farm information enters the village social system?

ii) Who to whom the farm information goes within the village social system?

Can be obtained from the findings of many research studies conducted in India and abroad. The two-step flow model of communication gives some clues to understand the process of farm information flow within a village social system. Van Dan Ban (1964); and Rogers and Shoemaker (1971) identified a number of deficiencies in the two-step flow model of communication and suggested few modifications. It is in the light of their suggestions and keeping, In view the findings of previous research studies, a diagrammatic presentation of the farm information flow process has been attempted (Fig.1.0).

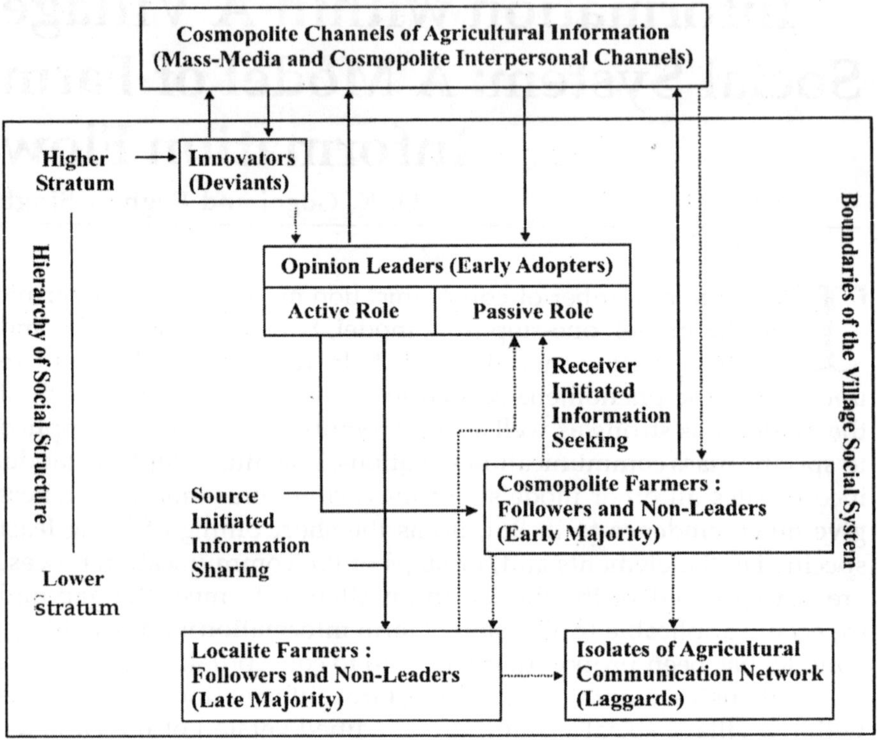

——→ Represents flow of farm information with high probability.

- - - -→ Represents flow of farm information with low probability.

·········→ Represents flow of farm information with very low probability. The upward arrow indicates receives initiated communication and the downward arrow indicates source initiated communication.

Fig. 1.0. Communication of farm information within a village social system : A model of farm information flow.

In the present model, the farmers of a village social system have been divided into five groups following the adopter categories (the normal distribution curve on the basis of innovativeness, Rogers and Shoemaker, 1971) and their communication behaviour. The first group is the innovators who are considered as "deviants" and relatively less integrated within the social system. The second group consists of the opinion leaders who are more integrated within social system and serve as role models for many other farmers of the village. This group is labeled as "early adopters" in adoption category curve. The potential adopters consider them as the "man to check with". The farmers (other than the opinion leaders) who may also be cosmopolites and exposed to mass-media but have little or no influence in others' innovation-decision process, are placed in the third category. They are labeled as "early majority" in adopter categorization. This group of farmers frequently interacts with peers and serves as linkages between early adopters and the late majority group (Rogers, 1983). Thus, they play an important role in the diffusion of farm information across social strata of the of the village social system. The fourth group consists of the farmers who are Localities with less or no exposure to mass-media and no extension contacts. They mostly rely on their opinion leaders or fellow farmers for information and advice on matters of farm problems. This group closely resembles the "late majority". The proportions of cosmopolite and localite farmers (the third and fourth category) will largely depend upon progressiveness of the village. The distribution in these two categories may not completely overlap 'with the percentages in the normal distribution curve of adopter categories. The laggards who are near "isolates" (Roger and Shoemaker, 1971) have been placed in the fifth category. The groups/categories have been placed in the diagram on a low to high social status continnum in the light of the previous research findings on socio-economic characteristics of the adopter categories. The late majority and the laggards (isolates) have been placed on the same plane of lower social stratum.

(i) How farm Information enters a village social system

The information on new improved farm practices enters the village social system through cosmopolite and progressive farmers who have access to the extension agencies and exposed to mass-media and then diffused through interpersonal channels (Singh, 1965; Shankariah, 1969; Rogers and Shoemaker, 1971; Singh, 1973 and Appa Rao, 1975). In the two-step flow model, it is stated that the opinion leaders directly get information from mass-media channels like radio, television, newspapers, magazines, etc. But later, Rogers with Svenning (1969) found that in less developed countries where little or no mass-media are available, other channels like personal trip to cities, conversation with change agents, etc. may provide information to opinion leaders. The findings of researches in India (Dhaliwal, 1963; Grewal, 1976; Singh, 1980 and Prasad, 1973) have similarly shown that it is not the mass-media alone, but the cosmopolite interpersonal channels which may also provide Information on new improved farm practices to the members of a village social system. So, the label "Cosmopolite channels" has been used to include the group of mass-media and the cosmopolite inter-personal channels.

The two-step flow model assumed that opinion leaders are the active individuals in the information seeking and information transmission process and the remaining mass audience are passive individuals. But the findings of Sicnski (1963); Write and Cantor (1967); and Rogers with Svenning (1969) indicated that either leaders or their followers are capable of active and/or passive roles in the flow of information. The active opinion leaders are those who seek information on improved farm practices from cosmopolite sources and initiate the information transmission process (through information sharing[1] behaviour) within the village social system. And the passive opinion leaders are those who seek farm information from outside sources, but take little or no interest in "self-initiated information transmission. In the two-step flow model, it is stated that the opinion leaders are the first to get the information from cosmopolite channels. There may be other groups of individuals who may also be the "first receivers" of any new farm information. One group may be the innovators who are venturous and more cosmopolite than the early adopters (Rogers and Shoemaker, 1971) and have the chance of receiving the new farm information earlier than others. The other group, the follower farmers (other than the opinion leaders) who are cosmopolite and exposed to mass-media and have extension contacts may also be the first to receive any farm information. The latter may be the case when the village social system is more progressive.

There is controversy over the innovativeness and opinion leadership overlap. Rogers and Shoemaker (1971) cited the findings of Menzal and Katz (1955); and Rogers and Burdge (1962) in support of the overlap thesis whereas the findings of Coleman and others (1966); Wilkening (1952); and Lionberger (1953) found little or no overlap. A partial solution to the controversy is provided by considering the effect of norms of the social system on innovativeness of opinion leaders. Rogers and Shoemaker (1971) resolve the controversy by drawing a generalization : "When the system's norms favour change, opinion leaders are more innovative but when the norms are traditional, opinion leaders are not especially innovative". Van dan Ban (1964) also concluded that the relationship of innovators and opinion leadership is much closer in progressive than in traditional groups. Sen (1969) found that leaders in more modern villages. So, it may be assumed that in more traditional villages. So, it may be assumed that in a progressive village, opinion leaders are the innovators whereas in a non-progressive village opinion leaders and innovators are separate individuals.

(ii) Who to whom the farm information goes within a village

When the information is received by innovators (if they are separate individuals from the opinion leaders), there is little probability that it will be passed on to other farmers of the village social system. But it may be diffused within their friendship circle or clique and many of their friends, may be outside the social system (Rogers and Shoemaker, 1971). As soon as the information is received by opinion leaders, there is high probability of diffusing the same among other farmers of the village. As a reaction to the information received, the first hand receivers may prepare themselves for innovation-decision processand/or pass on the same to other farmers with whom they interact. The information transmission may be the initiative of the source-person or initiated by the receivers. Rogers with Svenning (1969) labeled these two dimensions of opinion transmission as "opinion giving" (when the opinion leader himself initiated the process) and "Opinion seeking" (when follower is the initiator). It is unlikely that opinion leaders search their receivers and initiate the process of transmission. The most likely occurrence is that of "opinion seeking". Sen (1969) cited a number of research studies, in which it was found that "for advice on problems accuring from day-to-day life, people seek out others and are influenced by them". In the context of Indian village situations, social norms are such that leaders are not expected to go round the village informing people about latest technology or programmes. In a sense, interpersonal communication between leaders and followers is expected to be unidirectional. Unless people ask, leaders may not transfer knowledge. Even if leaders do not take the first initiative to communicate the latest knowledge to their followers, their leadership position is not affected (Gaikwad and Tripathy, 1974).

If the new information is received by cosmopolite farmers (the third group) they may approach their opinion leaders or friends either to know more or to get approval before adopting it. There are also research evidences that people receiving new information from interpersonal channels may go back to mass-media or other cosmopolite channels for more information. This may be the case for early adopters i.e. opinion leaders who are more cosmopolite and have higher exposure to mass-media than their followers. The follower farmers usually take back the information to their opinion leaders or progressive farmers to get approval of the new idea or to remove dissonant states if the information is inconsistent with their attitudes. Trol-dahl (1966) proposes on the basis of "balance theory" that followers who are exposed to mass-media messages that are inconsistent with their predispositions will initiate interpersonal communication with opinion leaders to reduce their dissonance.

The fourth group of localite farmers who are rarely exposed to mass-media or other cosmopolite channels have to depend on their opinion leaders for information and advice. As has been discussed, the flow of farm information from opinion leaders to followers of this group occurs through the information and advice seeking process. A substantial proportion of farmers of a village belongs to this category and many a times farm information reached them late because of their low level of information seeking behaviour. This may be the case in a non-progressive village. But in .progressive villages, it is likely that the localite farmers search out information within their village social system. The isolates who usually interact within their group, having traditional values, have little chances to get the new information in time. Their innovation- decision is very slow and remains lagging farm behind awareness-knowledge of a new idea (Rogers, 1983).

The flow of farm information from one group or individual to the other of a village social system is not a smooth process. There may be a number of factors like social structural barriers, interaction pattern (Homophilic-heterophilic) of the farmers, geographical locations, and progressiveness of the village. The information seeking and information sharing behaviour of the farmers may also have a direct effect on the flow of farm information within a village social system.

Many of the hypotheses proposed in the model will have to be tested in future researches. The model is a result of a theoretical exercise based on past researches and available literatures on diffusion of innovations.

REFERENCES

1. Appa Ran, G., "Structural and Functional Analysis of Interpersonal Communication Network in Diffusion of An Agricultural Innovation in An Innovative and A Non-Innovative Village In East Godavari District of Andhra Pradesh" Unpublished Ph.D. Dissertation, Indian Agricultural Research Institute, New Delhi,1975.
2. Coleman, James et at., "Medical Innovation ; A Diffusion Study, "New York : Bobbs Merrill, Citing Rogers and hoemaker(1971), Op.cit.
3. Dhaliwal, A.J.S., "A Study of some Important Factors Affecting the Adoption of a FEw Selected Agricultural Practices by the Cultivators of Ludhiana Community Development Block". Unpublished M.Sc. Dissertation, P.A.U., Ludhiana, 1963.
4. Gaikwad, V. R. and B.L.Tripathy, "Social Interaction and Communication in an Indian Village" Ahmedabad, Indian Institute of Management, P.14, 1974.
5. Grewal, I. S., "Multivariate Analysis of Adoption of High Yielding Wheat Technology in Arid, Central and wet zones of Punjab State", Unpublished Ph.D. Dissertation, P.A.U., Ludhiana, 1976.
6. Lazarsfeld, P.F. et. al. "The People's Choice" New York: Columbia University Press, 1944.
7. Lionberger, Herbert F., "Some Characteristics of Farm Operators Sought as Source of Farm Information in a Missouri Community", Rural Sociology, 18, PP. 327-338, 1953.
8. Menzal, Herbert and Katz, Elihu, "Social Relations and Information in the Medical Profession - The Epidemiology of a New Drug" Public Opinion Quarterly, 19, PP. 337 - 352, 1955.
9. Prasad, Rajinder, "Differential Contribution of some Correlates of Village Level Workers Influencing their Communication Effectiveness in the context of High Yielding Varieties Programme; A Study Conducted in the Punjab State". Unpublished Ph.D. Dissertation PAU, Ludhiana, 1973.
10. Rogers, E.M., "Diffusion of Innovations", New York, The Free Press, 1983.
11. Rogers, E. M., and F. F. Shoem aker, "Communication of Innovations : A Cross Cultural Approach". New York, The Free Press, 1971.
12. Rogers, E. M. and Lynne Svenning, "Modernization among Peasants : The Impact of Communication, New York, Halt, Rinehart and Winston, 1969.
13. Rogers, E. M. and R. J. Burdge, "Community Norms, Opinion Leadership and Innovativeness Among Truck Growers", Wooster,

Ohio Agril.Exp. Stat. Res. Bul. 912. Citing Rogers and Shoemaker (1971) Op. cit.

14. Sen, Lalit K., "Opinion Leadership in India, A Study of Inter-personal Communication" Hyderabad, NICD, 1969.
15. Shankaraiah, Ch., "A Study of Differential Communication Patterns in Progressive and Non-progressive villages, Unpublished Ph.D.Dissertation, 1ARI, New Delhi, 1969.
16. Sicnski, Andzezi, "A Two-Step Flow of Communication : Verification of an Hypothesis in Poland", Polish Soc. Bul. 1, PP.33-40, 1963, Citing Rogers and Shoemaker (1971) op. cit.
17. Singh, Tar loch an, "A Multivariate Analysis of Adoption of Improved Agricultural Technology of Rice Crop in the Punjab", Unpublished Ph.D. Dissertation, PAU, Ludhiana, 1980.
18. Singh, Y. P., "Key Communicators of Agricultural Innovations", Agra (UP), Satish Book Enterprise, 1973.
19. Singh, Y. P. "A Study of Communication Networks in Sequential Adoption and Key Communicators", Unpublished Ph.D. Dissertation, lARl, New Delhi, 1965.
20. Troldahl, V. C., "A Field Test of a Modified Two-Step Flow of Communication Model" Public Opinion Quarterly, 30, PP. 6G9-623, 1966.
21. Van dan Ban, A. W., "A Revision of the Two-Step Flow of Communication Hypothesis" in Lee Richardson (ed.), "Dimension of Communication", New York, Appleton-Century-Crofts, PP.263-274, 1964.
22. Wilkening, E. A. "Informal Leaders and Innovators in Farm Practices", Rural Sociology, 17, PP.272-275,1952
23. Write, C. R. and Cantor, M., "The Opinion Seeker and A voider : Steps Beyond the Opinion Leader Concept," Pacific Soc. Rev., 10, PP.33-43, 1967, Citing Rogers and Shoemaker (1971), Op. cit.

Chapter **17**

The Potential of the Mass Media as a Vehicle in Transferring Agricultural Technology to Farmers

George S. Atkins

The mass media in all its facets can, in no way, be considered in isolation as a vehicle in the transfer process. To be truly effective, its own components must be integrated, at all levels, with a carefully planned and executed agricultural extension programme. Then on an even broader scale, all aspects of this fully integrated programme should dovetail with a comprehensive "new order" for the basic industry of agriculture in a particular country a region or for the continent as a whole.

There are many types of vehicles on the roads in Africa. As I came here from Lagos, I saw lorries and pickup trucks, vans, buses, cars and land rovers - all vehicles involved in transferring people, goods and services from one place to another.

Just as there are various types of transport vehicles, there's a whole range of vehicles of mass media - so let's take a look now at the potential they hold in the process of transferring agricultural technology.

First, let me reproduce some extracts of an interview recoded away back in 1957. These are the words of B.R. Sen, who was, at that time, Director General of the Food and Agricultural Organization (FAO).

"The world at present does not produce adequate food for all its population. The basic fact is that there is not enough food in the world today."

That was more than ¼ of a century ago when the world's population was less than ½ of what it is now.

And what's the situation right here on this continent today?

Food production is not keeping up with population growth and we are told that the main reason is problems with the soil; there are, of course, many others as well. Because of them food production has fallen 11% per persons since 1970. The total food deficit this year is 5,000,000 tons – twice as much as it was decade ago.

The situation must be reversed; effective measures must be adopted; not one measure, but many – and in the right combination.

Indeed, there seems to be a new resolve in Africa as enunciated only recently in the so-called "Harare Declaration". In it a number of African governments pledged themselves to tackle the food problems once and for all. It's essential, however, that the governments be serious in making such a commitment, because examples can be cited of highly publicized national programs for which expenditure allocations have been woefully inadequate.

For 16 years, Colin Fraser has headed up Development Support Communication at FAO. He decries the fact that throughout Africa mass media, mainly controlled by governments, pay small attention to agriculture. One country he particularly mentioned to me, with 90% of its population living in rural areas, only devotes 1.2% of its broadcast time to rural programming. Colin told me only last month, and I quote, "It is useless to talk about the role of mass media in food production unless the political will to use the media from that purpose is forthcoming" end quote.

Only 3 months earlier, Soedjatmoko, Rector ofthe UN University in Tokyo, made his thoughts known on the Information Revolution. Among other things, he said that the traditional ways of transferring knowledge are "increasingly inadequate". He suggested that farmers and other on the periphery need their own radio stations but, he said, and I quote, "This will take political courage on the part of governments; but unless they take ... steps to create micro information environments, their development efforts are bound to remain ineffective".

Well, let's assume that in the new order here in Africa political will and resources will allow for the development of the mass media as a vehicle for the transfer of agricultural technology.

That is it we need to transfer?

Dr. Daniel Benor is one of the developing world's best known experts in agricultural technology transfer. Back in 1978, World Bank

President, Robert McNamara considered him "a genius". I met Benor in Washington and we talked about his formula for increasing crop fields by 50 to 100%. The way he put it was "we just close the gap between what is know and what farmers do". So our job at this conference is to passes the potential of mass media in this process.

Now, in the past, many have thought that, by simply selling the farmer about an improved method, he or she would adopt it. We know, of course, that his seldom happens. We are also aware that "to close the gap between what is known and what farmers do". Early always takes more than just one kind of approach. Colin Fraser says, "The physical presence of a good extension worker, or more advanced farmer, is bound to be necessary some stage of the adoption process. Nevertheless, less media, says Colin, "Do have a vital role to day in informing farmers that innovations exist, and in motivating them to seek further information out them."

Personally was a full-time farmers in Canada for 15 years and, from my own experience and that of my neighbours, as well as from my own observations developing countries, I know this is true.

The best results are achieved by an integrated approach thereby all the mass media, and the communicator who acts the farmer face to face, are telling the same.

The mention 3 or 4 examples, all of which are associated in various ways with the developing countries from Radio Network (DCFRN). Two are here in Africa there is another in the Indian Ocean and one in South-East Asia.

In Malawi, there's a three-pronged approach, using extension agent, training centres and mass media.

The Extension Aida branch of the Ministry of Agriculture handles the media work with a staff of 100 people working on radio programmes, printed material, a cinema van service, evaluation and research.

Within that mass media element, many different types of vehicles are used and with a considerable amount of imagination on the part of the sizeable staff. A visit to Malawi would convince you, I am sure, of the value of the system they use.

In Swaziland there is a national media production centre and they produce printed publications, graphics and audio-visual material, including videotapes for use around the countryside.

Moving east, Seychelles of course is a very small country but there, all communicators are telling the same story at the same time - on radio on videotapes shown regularly in community centers, at field days and in printed material handed out wherever farmers gather. In addition, the information always appears once a week in a farming section of The Nation, the country's daily newspaper.

The most ambitious integrated programme I know of has been established for more than a decade in the Philippines. It's called "Masagana 99", - that means "Bountiful Harves", and the number refers to the programme's target yield of cavans per hectare. This one has 11 elements ranging from a research based technology package, through a farm input allocation and distribution system, to a major mass media component - and incidentally, as radio reaches nearly all Filipino farmers, this vehicle figures very prominently in the dissemination of information.

All who are in touch with farmers by radio are involved "Masagana 99" in other ways as well and they also attend training sessions where they regularly work at improving their communication skills. The principal backup media vehicles are in the print category, including carefully prepared instructional comic book style publications, periodical, newsletters, booklets and bulletins in the eight major Filipino dialects. There are also posters and the full support to newspapers in all 59 provinces.

"Masagana 99" is credited with having changed the status of the Philippines from that of a rice importing country to oneof the world's few exporters or rice.

Now for a few minutes, let us consider some of the vehicles of mass media communication. To start with, let's think about the potential for radio of all the mass media, here is the one with the greatest potential. There are two main reasons :

The first is the increasingly large numbers of the African people who can be reached by radio. Four years ago there were 18,000,000 radios in Africa and that number has now increased to 29,700,000. This is one radio for every 16 people. Mind you, it's still not perfect because batteries run down and maybe unavailable and radios become inoperative for other reasons, but radio is indeed a formidable communication tool.

The second reason, radio has the greatest potential in very large numbers of the target group of farmers to be reached are illiterate.

They may not understand the printed word; indeed, they may not even absorb the message from pictures- but the spoken word, they know; - and if it is used skillfully, communication with or among them can be complete and effective.

While television is on the increase, it's not yet realistic to consider this as a viable vehicle for reaching vast numbers of farming people.

Now, there are many who believe that when communicating with people who can not read or write, pictures will convey the desired message. This is not necessarily so and a number of studies in developing countries have proved it.

Of course, we, who consider ourselves literate, would easily recognize a drawing of nothing more than a hand; but when such a drawing was shown to a test group of rural adults in Kenya, only 69% of them could identify what they were looking at. Similarly, only 31% identified a line drawing of a cob of maize and as few as 10% knew what a drawing of a common housefly was supposed to represent.

Surely, when we are considering the potential use of pictures for transferring information to farmers, studies like these must signify that great care, collaboration with experts and pre-testing must be included in our planning. This should apply whether we are thinking of posters, filmstrips, slide/tape presentations, or video taps. Indeed the same principle should be applied no matter what mass media vehicles we plan to use.

Our target audience in various parts of this continent will vary considerably in their preconditioning due to cultural differences — in fact, the vehicles we use must be uniquely adapted to suit those farmers whom we wish to influence.

In a moment, we'll consider some aspects of the content of the message these various vehicles might carry. First, though, to give you a sense of the great diversity of vehicles that can be used in a single medium, let me just list a few of them in radio broadcasting.

Straight radio talk or commentary. (For variety, it can include more than one person sharing the spotlight.)

- Live or recorded interview (in the studio or on location on the farm, in the cattle shed, in an office or laboratory, etc.; or it might be a live or recorded telephone interview).
- Mailbag format (questions and answers).

- Drama - (scripted radio play complete with sound effects).
- Unscripted dialogue (could include extension agent and farm family members).
- Character impersonation (old woman or old man of the village conversing with extension agent, etc.).
- Documentary.
- Live or recorded remote location broadcast(entire programme).
- School-on-the-art.
- "Commercial" type spot announcements.
- Jingles and/or slogans.
- Farm songs.

Closely allied with all these "spoken word" vehicles are the many techniques that can be used in group get-togethers with farmers. There's the whole gamut of flip charts, posters, and the like, as well as cassette tapes that have been carefully prepared for presentation in a group setting. All these are, in effect, mass media vehicles, are the lectures, training schools, short course, workshops, field days, demonstrations and group tours.

One more mass media category I would add that is particularly effective for farmers who have some mobility is the kind of facility that UNICEF set up a number of years ago in Nairobi, the so-called Village Technology Unit. In it, various types of technology can be demonstrated on an ongoing basis. This technique has been widely developed under the name "Self-Reliant Home Movement" in the Philippines, with a nation-wide network of half-acre demonstration homesteads. I recently visited one of them northwest of Manila, and I found it to be a very practical and relatively inexpensive mass media vehicle. (Incidentally, while I was there, I recorded information of composting, several innovative cropping practices and even simple biogas production. It will be distributed to communicators in more than 100 countries who are participants in the Developing Countries Farm Radio Network).

One of the great strengths, of course, of the types of vehicles that bring farmers together in the participation and two-way communication that can take place between the communicators and the farmers and among the farmers themselves.

I must mention here that, with the development of so-called community radio stations and perhaps those special farmer radio stations the Rector of the UN University is advocating, opportunities for much more of this two-way communication can become possible, even in radio programming.

The quality and effectiveness of all these vehicles in the process of technology transfer will be directly proportional to the adaptability skill and training of the communicator(s) involved and the infrastructure that is in place to support them.

These mass media vehicles can be expected to "close the gap between what is know and what the farmer does".

On Tuesday of next week (October 16th) World Food Day will be universally observed. Three years ago, in press information kits that were widely distributed, FAO highlighted the fact that in the past, the great majority of the world's food producers – those small village level subsistence farmers, had been taken for granted. In fact, their potential for increasing world food supplies had been largely ignored by the big multi-million dollar development schemes and projects. FAO's conclusion was that these farmers must now play a central role in efforts to overcome hunger.

This makes real sense. So let us target our efforts in this direction. As we do this, I believe the transfer of technology should not just be geared to "increasing food production"; I prefer to use the phrase "increasing food supplies".

This includes :

- Increasing food production.
- Decreasing losses of food that has been produced.
- More efficient use of food - and this, of course, includes information related to nutrition and health.

Just think for a moment about those hundreds of millions of farmers in this target group and consider that if 25%, or even only 10% of them were to increase food supplies in their respective venues to any meaningful extent, what a significant increase that would be.

This, however, brings into sharp focus the fact that if any really serious attempt is begin made to increase food supplies by whatever means along with it must go a very carefully conceived component aimed at marketing and distribution. Let us not repeat past mistakes of encouraging farmers to produce food they have no way of selling.

By the same token, we must be careful not to motivate farmers toward some action that they can not possibly carry out.

For example, let's say a farmer has been persuaded that increased yields can be obtained from legume plants that have grown from seed inoculated with a new strain of Rhizobium Bacteria. Nothing but frustration will result if that farmer is unable to obtain this new input. Indeed, Dr. M. S. Swaminathan, Director General of the International Rice Research Institute, told me earlier this year that this is one of the quickest ways for a rural communicator to destroy his or her credibility.

What I am really saying is that the technology being transferred must be relevant to the farmer who receives it.

It must also be culturally acceptable. I well recall the story of a village poultry improvement project that failed because the new hens that were introduced were white. The people already had white goats and it was their belief that two different kinds of domestic animals of the same colour could not possibly be compatible. The project would have been successful if the introduced hens had been brown. Here again is an example of the need for a multi-disciplinary approach to all facets of the transfer of technology.

And now I come to a crucial point regarding the target group. In the past nearly all, if not all, communicators in this field though of farmers as men only. What an incredible mistake! Anyone who gives the question any serious thought will realize that, throughout Africa, women are, in fact, more involved with food production than men. It is also true that, in many instances, it is they who make family and farming decisions and, in may others, they have a very definite influence on decisions that are made. If our attempts to "close the gap" are to be successful, we must see to it that a major change in this direction comes about. It must come about in the way every vehicle of communication is used.

In addition to what I have just said, we are also missing out if we ignore the younger family members who will be the farmers tomorrow. There are many good reasons why mass media in the rural context should be adjusted toinclude the millions who comprise this segment of farming people.

Now, just a word on the mass media communicators who drive and service all these "vehicles" I have been talking about. Should we start with communicators and give them extensive training in agriculture, or should we do it the other way around?

Jan Kinket, the chief training officer for Radio Nederland, has trained hundreds in courses right here in Africa. He told me he is convinced that the best way is to train a good agricultural extension worker to become a broadcaster.

The communicator must have an abiding interest and any empathy with farming people. To be most effective, this person must consider him or herself to be a servant of the farmer; not some exalted person who talks down from on high, dispensing information only because that's what he or she is being paid to do.

References :

1. Sen, B. R., Interviewed by George Atkins for broadcast on the Trans Canada Network of the Canadian Broadcasting Corporation, May 17, 1958.
2. Critchfied, Richard., "Africa's Food Crisis Has Deep Centuries - Old Roots", The Christian Science Monitor, June 6, 1984.
3. Daddieh, Cyril., "Self Sufficiency Destroyed" - African Food Crisis Series - 1, Development Forum published by the UN Division for Economic and Social Information/DPI and the United Nations University, Tokyo, Japan,, July-August, 1984.
4. Fraser, Colin., Private correspondence between Colin Fraser and George Atkins, dated August, 13 and September 7, 1984.
5. Soedjatmoko., "South Must Play Active Role in Information Revolution, United Nations University Newsletter, Tokyo, Japan, Vol. 8, No. 1, June, 1984.
6. Rowen, Hobart., "Making More Corn Grow:, The Guardian January 8, 1978.
7. Benor, Daniel., Intervіwed by George Atkins for Massey Ferguson Climate/Food Project in Washington, D.C., USA, May 7, 1978.
8. Onyango, Otieno Mak., "African Nation see continual co-operations as means or enhancing broadcast development", Television/Radio Age International June, 1980.
9. Lynn, Wilma., "Radio Coverage Statistics", Development Communications Report, Washington, D.C., USA, No. 46, summer, 1984.
10. Jarmul, David., Plain Talk (Clear Communication for International Development) Published by Volunteers in Technical Assistance, 3706 Rhode Island Avenue, Mt. Rainlier, Maryaland 20822, USA.
11. Fussell, Diana and Ane Halland., Communicating with Pictures in

Nepal, National Development Service, Tribhuvan University, Kathmandu, Nepal and UNICEF Lazimpath P.O. Box 1187, Kathmandu, Nepal.

12. Self Relian Home Primer/Overview, 1984. Ministry of Human Settlements, Quezon City, Manila, Philippines.
13. Swaminathan, M. S. Interviewed by George Alkins for the Developing Countries Farm Radio Network at the International Rice Research Institute, Los Banos, Philippines, July 24, 1984.
14. Kinket, Interviewed by George Alkins for Massey – Ferguson Climate/ Food Project in Hilversum, Netherlands, November 24, 1978.

Chapter **18**

Mass Media and the Psychology of the Farmers

N. B. Chauhan

The present age has been rightly termed as an "information age". People want adequate and authentic information as early as possible. Farmers as human beings are also anxious and become more desirous with the advancement in science and technology to know what is happening in the field of agriculture. They have eagerness to obtain knowledge, particularly in the field of modem agriculture to make themselves psychologically strong and conductive to adopt modern methods of agriculture. In India, it is difficult to contact each and every farmer in limited time in an effective manner to transfer agricultural technology. The use of various mass media is certainly most effective avenue to convey information to the broad means of people, particularly to the huge illiterate segment of the farmers. In Gujarat, farmers have been communicating agricultural information through farm television, farm radio and newspapers to create conductive behavioural changes among them. Looking into this fact, present study on impact of mass media on the psychological performance of the farmers was undertaken.

Methodology

The study was undertaken in randomly selected twelve villages of Anand and Kheda districts of Gujarat state on a random sample of 120 farmers. The data were collected through personal contact method and analyzed by using suitable statistical tools like percentage, mean standard deviation and coefficient of correlation. The scales developed by different social scientists were used to measure various behavioural changes in the farmers.

RESULTS AND DISCUSSION

Mass media exposure of the farmers : To know mass media exposure of the farmers scale adopted by Kamat (1993) was used with some modifications. After getting score of mass media exposure of each farmer, all the farmers, were categorized into two groups based on mean score. The farmers below mean score and above mean score were considered as farmers with favourable mass media exposure and farmers with unfavourable mass media exposure, respectively. It can be seen from table 1 that slightly more than half (55.00 per cent) of the farmers had favourable mass media exposure. It shows that majority of the farmers had considered mass media as important means of collecting useful information of modern agriculture.

Table-1 Distribution of the respondents according to their mass media exposure.

S. No.	Mass media exposure of the farmers	Number	Percentage
1.	Farmers with unfavourable mass media exposure	54	45.00
2.	Farmers with favourable mass media exposure	66	55.00

Table-2 Relationship of mass media exposure with different Psychological changes of the farmers.

S. No.	Variables of the farmers	'r' value for the farmers with favourable farm television exposure, N=66
1.	Level of aspiration	0.204 NS
2.	Educational aspiration for son	-0.109 NS
3.	Role taking empathy	-0.092 NS
4.	Non-fatalism	-0.021 NS
5.	Cosmopoliteness	0.041 NS
6.	Economic motivation	0.017 NS
7.	Risk orientation	0.231 NS
8.	Scientific orientation	0.495*
9.	Market orientation	0.221 NS
10.	Credit orientation	0.441*
11.	Innovation proneness	0.405*
12.	Attitude of modern agriculture	0.390*
13.	Adoption of modern agriculture	0.454*
14.	Over all modernization	0.442*

*Significant at 0.05 level of probability.

Psychological changes in farmers through mass media : The data presented in table 2 indicate relationship between the mass media exposure of those farmers who has favourable mass media exposure and level of various psychological changes taken place among them.

1. **Mass media exposure and level of aspiration** : The data in table 2 state that level of aspiration of the fanners was not observed significant with their level of exposure of mass media. In other words, it can be said that mass media did not play significant role in improving degree of expected rational estimation of the farmers to reach up to certain level of development for future.
2. **Mass media exposure and level of educational aspiration for son** : Table 2 depicts that farmers with favourable mass media exposures' level of aspiration for their son was not improved with increase in their level of exposure of mass media. In other terms, it can be concluded that through mass media, farmers have yet not understood that high level of education to new generation is essential for their all round development. Mass media had not made farmers with such a realization that whatever difficulties they have faced due to lack of their expected level of education should not be repeated for the generation.
3. **Mass media exposure and level of role taking empathy** : The mass media exposure had not shown its positive role in improving ability of the farmers to take and perform role of other persons and not educated them to be a flexible and adjustable to tackle any situation by using a skill of being a part of other than their own occupations. Such kind of skill is very useful to the farmers at the time of failure of the agriculture to sustain their families in difficult situations, therefore to develop such skill mass media should concentrate their efforts.
4. **Mass media exposure and level non fatalism** : The relationship between mass media exposure of the farmers and their degree of non fatalism was observed non significant indicating that mass media did not show its impact to convert fatalistic fanner into non-fatalistic one.
5. **Mass media exposure and level of cosmopoliteness** : The mass media exposure showed non significant impact on the cosmopoliteness of the farmers. The reason being the farmers might have fulfilled their curiosity about external world with increase in exposure of mass media.

6. **Mass media exposure and level of economic motivation** : The result showed that level of economic motivation of the farmers was not increased up to the level of expectation among those farmers who had favourable exposure of mass media. Here, it can be said that mass media have yet not developed necessary tendency and inclination among the farmers leading towards high level economical end.

7. **Mass media exposure and level of risk orientation** : The mass media exposure was not significantly correlated with the risk orientation of the farmers. It means that mass media exposure did not play noteworthy role to make farmer with a mind to take risk and courage to face uncertainty in doing something remarkable to improve his existing situation.

8. **Mass media exposure and level of scientific orientation** : The data in table 2 indicated that mass media exposure played significant function on the scientific orientation of the farmers. It means dial mass media has done something to make farmers with a mind to use science in decision making in relation to their adoption behaviour.

9. **Mass media exposure and level of market orientation** : The mass media exposure was non significantly correlated with the market orientation of the farmers. It proves that mass media had not played useful role to improve degree of knowledge of farmers in terms of market. It has yet to do something extraordinary to make farmers able, clever, talented, capable and competent to take judgment to sell their products for better price by analyzing various prevailing infrastructureand market intelligential.

10. **Mass media exposure and level of credit orientation of the farmers** : The results in table 2 pointed out that mass media exposure played significant role to make farmers with accredit borrowing nature and able to take advantages of loans given by various agencies. It means that mass media had done amazing and inspiring job to craft farmers with a high level of understanding and credit borrowing behavioural capacity to make best use of modern high production oriented inputs.

11. **Mass media exposure and level of innovation proneness** : The data in the table 2 disclose that mass media played crucial responsibility in improving level of innovation proneness of the farmers. The findings support the general view that through

high degree of exposure of mass media person becomes knowledgeable and innovative in several aspects of new technology.

12. **Mass media and level of attitude of the farmers towards modern agriculture** : As far as impact of mass media on attitude of the farmers towards modern agriculture was concerned, it was observed significant Mass media like Farm TV being an audio-visual means of communication might have presented some programmes on the results of successful farmers who have got advantages of modern practices of agriculture. The other media like radio and newspaper might have also done something excellent to change the attitude of the farmers towards modern methods of farming by comparing results of modern technology with traditional one through giving chance to the successful farmers to tell their experience on radio and newspapers.

13. **Mass media and level of knowledge and adoption of modern agriculture** : The level of knowledge and adoption were observed significantly superior among those farmers who had high extent of exposure of mass media. It means that mass media did a lot to improve knowledge and adoption of the farmers regarding modern practices of agriculture. Thus, it is worth to say that mass media are valuable means to convey information to the broad means of fanners for its effective adoption.

14. **Mass media and level of overall modernization** : As far as level of modernization of farmers was concerned, mass media played considerable role in improving it. The greater contact with the large society via media leads persons to develop different mental abilities and capacities to deal with consequences of modernization. This might be the reason for above finding.

CONCLUSION AND IMPLICATIONS

From the above study, it can be concluded that majority of farmers had favourable exposure of mass media. As far as behavioural changes in farmers were concerned, favourable mass media exposure had shown valuable impact in improving level of scientific orientation, credit orientation, innovation proneness, attitude towards modern agriculture, overall modernization and knowledge as well as adoption of modern practices of agriculture of the farmers. Whereas, mass media did not show its impact to improve farmers' level of aspiration, educational aspiration for son, non-fatalism, cosmopoliteness, economic motivation, risk orientation and market orientation. It is therefore, recommended that mass media should also concentrate their efforts for those areas of behaviour of farmers which are yet to be changed. For this, more participation of farmers should be encouraged by those persons who are directly concerned with development, planning, implementation and execution of various mass media at different stage, from the selection of content to final broadcasting, telecasting and presentation stage of various media.

REFERENCE

1. Kamat, V. (1993). Kheda District Communication Project, Evaluation Study : Unpublished report, S.N.D.T., Women's University, Bombay.

Chapter 19

Methodologies to Assess Media Effects and Utility

G. Perumal, K. P. J., Krishna Kumar, K, A. Ponnusamy

In societies with certain extent of developed media systems, according to Ball-Rokeach and De Fleur (1976), audience dependency on media information increases as the level of structural instability (social conflict and change) increases. Dependency otherwise means the potential of the media messages to achieve a broad range of cognitive, affective and behavioural effects on the part of audience. They are the effects that are not easily detected in laboratory experiments or in before after studies of people who have been exposed to specific messages. Nevertheless, efforts are being made constantly by the media researchers to measure the effect of media messages on their audience and the extent of utilization of media messages by them as objectively as Possible and also standardize such research methodologies. This paper aims to provide a summary of selected valid research methodologies used to measure the effectiveness of farm media such as Farm broadcast, farm telecast. Farm magazines and Farm pages of Newspapers in terms of their perceived effect and utilization by the farmer audience.

Setting and objective

The methodologies to measure the media effect and utilization are drawn from various media studies conducted in Tamil Nadu Agricultural University (TNAU) and summarized in this paper under the following heads :

(i) 'agenda setting effect' of the media and 'audience activity' in relation to the farm messages published in the Newspaper, farm broadcast and farm telecast.

(ii) 'media creditability' as perceived by the audience.

(iii) 'audience preferences' towards different farm journalism/ broadcast/telecast components.

Conceptual clarification and operationalisation of these methodologies as provided in this paper can evoke a debate among the media researchers in respect of compatibility and applicability of the methodologies over time and space, ultimately resulting in their standardization.

(I) Measurement of agenda setting effect and audience activity

The pioneer study on the "agenda setting effect of farm media" was conducted by Krishnakumar (1990) at TNAU. A new approach, called 'agenda setting', to study media effects was proposed in 1970, although scholars had written about it even earlier. In a way, it represents a more sophisticated understanding of the complexities of communication McCombins and Shaw (1972) first coined the word 'agenda setting' which meant the ability of the media to transmit to the audience the issues that will be included in Public debate. Murphy (1977) called the agenda setting function as the ability of the media to determine what we talk about, just by deciding what information to print or air. The central hypothesis of agenda setting, according to Becker *et. al.* (1975) was the casual relationship between emphasis given to issues by the media and saliency assigned to these issues by audience members. Ball Rokeach and Defleur (1976) explained agenda setting as an interactional process in which topics were selectively transmitted by the media to the audience, who then would sort out their interest and concern with this information as a function of both their individual differences in personal make up and their location in social strata and categories.

Based on the above theoretical orientation, the process of agenda setting could be summarised as "mutually influencing interactional process taking place at information source, media and audience levels." In the context of transfer of farm technology, farm messages are sorted by the source for presentation to the media; media in turn selectively communicate them to farmer audience who then sort out the messages based on their personal make up, needs and their position in the social system. The farm information agenda setting was studied in terms of 'source, salience's media and audience'. The information selectively supplied by the source to the media is termed as 'source salience'. The topics sorted by the media for presentation to the farmer audience are called as 'media salience'. Among these messages which are selectively perceived by the audience are 'audience salience'.

Study of source salience

Source salience refers to the salient messages which are sorted out by the source of information for communicating the audience through the media (Gumpert and Robert, 1986).

In the study reported (Krishna Kumar, 1990) the 'message' referred to the Farm information; 'source of information' refers to the State 'Department of Agriculture'; 'audience' refers to farmers and 'media' refers to Newspaper, Radio and Television.

The District level Deputy Directors of Agriculture (Information) and their Technical Assistants constituted as the respondents (45) for 'source of information'. Their selective perception of factors based on their importance in influencing source salience from among the 33 identified factors were obtained through mailed questionnaire.

The mean salience score for each factor was obtained as :

Most
considered
5 Frequency of responses x corresponding weights
1
Least
considered
Total number of respondents i. e. 45

Study of media salience

Media salience is the salient messages which are selectively assembled, selectively, determined and communicated to the audience by the 'media' (Gumpert and Robert, 1986).

Media do not transmit all the messages received from the source as such. Selection and editing (giving selective prominence) takes place at the media level to set an agenda among farmer-audience. Farm Radio Officers (FROs) of all the six AIR stations in Tamil Nadu, farm telecast programme producers of Doordarshan Kendra, Madras and Editors and Special correspondents of Tamil and English newspapers and some popular farm journals were the respondents (17) to study the factors influencing media salience. Similar procedure as described under source salience was followed to determine the 'mean media salience score' of each factor. In order to find out the correlation between media issue emphasis and subsequent audience issue salience the respondents were also asked to indicate the degree of prominence given to different farm subject matter of publication/broadcast/telecast.

Study of audience salience

Audience salience refers to the messages selectively perceived by the audience, specifically referred to the farmer audience in this study.

The study attempted to analyze the audience salience and by that assess the effect of agenda setting through audience activity. Audience activity is central to the agenda setting function of the media because the active audience either accept or resist media influence in an active way through their voluntaristic and selective orientation towards media messages (Klapper, 1960 and Levy and Windaul, 1985). The active participation by the audience in the communication process might be facilitated, limited or otherwise influenced by the agenda setting functions of the media and hence as Tipton *et. al.* (1975) pointed out 'audience activity' was considered as the crucial factor to the assessment of the agenda setting; effect of the media.

'Audience activity' towards media messages and is the recently introduced concept into the farm media research and is derived from the study reported by Levy (1983). The activities in relation to the farm messages of Newspaper, Radio and Television were suitably modified and combined by Krishna Kumar (1990) to study the comparative agenda setting effect of farm media.

Measuring audience activity

Fundamental to the study of media effects is the concept of an 'active' audience. The term 'audience activity' postulates a voluntary and selective orientation by audiences towards the media messages. Levy (1978) suggested that audience activity might be better conceptualized as a variable construct with audience exhibiting varying amount of activity.

Based on the typology of audience activity suggested by levy (1983), three types of audience activity viz. 'Pre-exposure selectivity', 'exposure involvement' and 'post-exposure utility' were studied by Krishna kumar (1990) in relation to the audiences exposure to the farm page of Newspaper, farm broadcast and farm telecast. Sample-audience for the study were drawn separately for farm page, 'farm broadcast and farm telecast from the members of active Farmer' a discussion groups (FDGL).

Pre-exposure Selectivity : Referred to the degree of intentional seeking to media messages/messages from farm page/farm broadcast/farm telecast Intentionality was meant to contrast with 'casual or accidental exposure' to media messages.

Exposure involvement - Referred to the degree of attention to media messages and engagement in information processing. Two measures of exposure involvement viz. 'attention' and 'para-social interaction' between audiences and media communicators was studied.

Pre-exposure utility : Referred to the extent of utilization of messages acquired from the media.

Developing tool to measure audience activity : The scale used by levy (1983) to measure the audience activity was considered as the basis to develop the tool consisting four sets of statements reflecting four-types of audience activity viz. Pre-exposure selectivity, attention, para-social interaction, and post exposure utility, consistent to the existing local situations and audience background. As suggested by Levy and windahl (1985) certain kinds of logical associations were also considered among activity types while formulating statements. For example highly selective exposure to media message is linked to active anticipation of exposure. Further active involvement would result in high gratification. For each type of activity three sets of statements separately for newspaper, Radio & Television were developed. Each set contained 10 statements. These statements were then validated by subjecting them to a panel of six judges who were asked to select those statements that could best reflect the type of audience activity it represented with respect to each medium. In total, 12 sets of statements (four types of audience activity x three media i.e. Newspaper, Radio and Television) were developed. The number of statements in each set, after the validation varied from five to eight.

The statements related to 'pre exposure selectivity' reflected one's deliberate intention to get exposed and efforts taken to not to miss the exposure to the farm page/farm broadcast/farm telecast. The statement related to 'attention' reflected concentration to the media messages and notes taking on marking salient points. 'Para-Social interaction' was reflected by the familiarity of the audiences with the media personnel, contact and closeness between them, mutual concern persuasive ability of the messages and extent of feedback. Audience activities like discussion with others, using media messages, sparing notes to others and clarification of doubts as put in the form of statements reflected, 'post exposure utility'.

The 'reliability' of all the 12 sets of statements was established through 'test-retest method'. Their content and construct validities were also ensured as suggested by Gareet (1969). The reliability co-efficient of 12 sets of statements varied from 0-78 to 0-90 and their validity coefficient varied from 0/89 to 0'95.

Measurement of audience activity

The audiences of the farm page of the Tamil Newspaper 'Dinamani', farm broadcast of AIR, Coimbatore and farm telecast of Madras Doordarshan Kendra were subjected to respond to each one of the statements on a three point continuum. Never sometimes and always. The points carried scores of 1, 2 and 3 for the positive statements and scoring was reversed for negative statements. The final score for each specific type of audience activity was obtained by summing up the score of the respective individual statements. The score for the type 'exposure involvement' was obtained by adding the scores of the two types viz. attention 'and para-social interaction'. The scores of all type of activities were added to arrive at the overall score for 'audience activity.'

The study of audience activity in relation to agenda function of the media assumes greater importance in the context of less active audience. Atkin (1973) reported that less active audience exposed themselves only to the media messages that were most prominently available. Tipton *et. al.* (1975) suggested that individuals who were less actively engaged in communication process. This suggestion is more appropriately applicable to Indian rural media with its majority of audience remain to be activated substantially.

Assessment of media credibility : Many media studies have adequately established that the variable; 'perceived media credibility' is crucial to build up favourable attitude towards media and its utility by the audience. Media credibility was found to be positively influencing audience activity and agenda setting effect of media by Krishnakumar (1990.) Most of the media researchers used paired comparison scale to assess the comparative media credibility between Radio, Television and Newspaper/Journals. In order to measure the media credibility more objectively, a 'semantic differential scale' was developed by Krishnakumar (1990) based on the scale designed by Berlo, Lement and Mertz as quoted by Joseph (1976).

The media credibility is largely a matter of relative faith of the audience in the media passages. The ten items of semantic differentials of this scale measure three dimensions on aspects of farm media credibility, viz. trustworthiness, expertness and personal dynamism over a seven point continuum.; these ten components were elected by a panel of ten experts based on their relevancy to credibility of farm media messages. The seven point continuum ranged from a Positive end associate with the highest credibility to the negative end associated with the lowest credibility. The specimen of the scale as administered to the farmer-audience (respondents) is given below:

In order to assess the extent of credibility, you attach to the farm page published in 'Dinamani'/Farm broadcast of AIR. Coimbatore/Farm telecast of Doordarshan, Madras, certain opinions of farm media messages are given at the left end. Just opposite opinions are given at the right end. The space between the two ends is divided into 7. Please indicate your position in the given contimuum by putting of mark at the appropriate point.

The score for each component ranged from 1 to 7 with the 'positive' and having the maximum score of 7 and the 'negative' end having the minimum score of 1. Thus, the total credibility score ranged from the minimum of 10 to the maximum of 70. The reliability and validity of the scale was established through test-retest method. The reliability coefficient was 0.8120 and that of validity was 0.9011.

Generally the information given in the farm page/farm broadcast farm telecast are								
Honest	7	6	5	4	3	2	1	Dishonest
Accurate	7	6	5	4	3	2	1	Inaccurate
Timely	7	6	5	4	3	2	1	Untimely
Authentic	7	6	5	4	3	2	1	Not authentic
Adequate	7	6	5	4	3	2	1	Inadequate
Useful	7	6	5	4	3	2	1	Useless
In suitable	7	6	5	4	3	2	1	In suitable
Language	7	6	5	4	3	2	1	Language
Compatible with audience situation	7	6	5	4	3	2	1	Un-compatible with audience situation
Reflecting experience in farm writing/broadcast /telecast	7	6	5	4	3	2	1	Reflecting experience in farm writing/broadcast /telecast
Active (motivating)	7	6	5	4	3	2	1	Passive

Assessment of media gratification

Research on uses and gratifications provided the greatest number of answers from individuals about why they select, what they do from the Kaleidoscope of media channel and content options (Donehew et. al. 1987). According to Lin (1977), the decision to enter into communication is motivated by goals and uses that are self-defined. Hence, the best way to measure media gratification is to ask the respondents as to what extent they are gratified in terms of satisfaction of their needs, through the media participation. Krishnakumar (1990) found that media gratification had significant and positive correlation with perceived media credibility, attitude towards media messages and audience activity. The perception of the audience about the extent of degree of utility of the farm page published in the Newspaper/farm broadcast/farm telecast was considered as the perceived media gratification in his study on farm-media agenda setting. He adopted a simple scoring procedure to measure the degree of media gratification :

Degree of gratification	Score
No gratification	0
Partial gratification	1
Full gratification	2

The gratification profiles of the audiences of farm page, farm broadcast and farm telecast were drawn separately based on the mean values and standard deviation of the gratification scores of the respective media groups.

Methodology to assess audience's preferences

Surveys to find out the audience preference and utilization of media messages are being carried out since as early as 1941. Audience surveys relation to printed farm messages in farm magazines and Newspapers, farm broadcast and telecast messages have been conducted in TNAU on following aspects :

The researches conducted in the area of 'audience preference' towards various media components, though not adequate has been comparatively more explored than other areas of media research. However, there is a scope for future research in terms of preferences of specific audience groups such as rural woman, farm youth etc. Advertisements, style of writing or presenting information and hardware aspects of media. Research findings in the above lines would help to strengthen the farm media research which is an emerging area in the field of agricultural extension.

Further, the increasing magnitude of coverage of audience by mass media with about 40% of the audience reached by television and even upto 90% by radio on one hand and the appreciable increase in the proportion of farm information being included in these media and the phenomenal growth of farm magazines, agricultural columns in newspapers especially at regional levels on the other hand warrant reorientation of the researches in a way that specific recommendation can emerge with regard to the audience needs, message treatment and the effect and utility of media.

References

1. Atkin, C. (1973) "Instrumental Utilities and information seeking." in P. Clarke (ed.) New methods for communication research Bevery Hills A. Sage.
2. Badrinarayan, P. (1977) "A study of farm broadcast listening behaviour of small farmers." unpublished M. Sc. thesis submitted to Tamil Nadu Agricultural University (TNAU), Coimbatore.
3. Balasubramanian, U. A. (1983) Relative effectiveness of cover page and format components of Valarum velanmai. Res. Report No. R. 1. Ag. Extn. 17. Dept. of Agri, Extension and Rural Sociology, TNAU, Coimbatore.
4. Ball-Rokeach and M. L. Defleur (1976) "A dependency model of mass media effects," Communication Research 3 : 3-21.
5. Beeker, L.B. M. E. Me. Combs and J.M. Me. Leod (1975). "The development of Political cognition" in S. H. Chafee (ed.) Political communication, Bevery Hills, C A : Sage.
6. Chandrakandan, K. (1982) "Effectiveness of farm broadcast on listeners affective. Cognitive and psycho-motor, behaviour" unpublished Ph.D. thesis submitted to TNAU, Coimbatore.
7. Donehew, L. P. Palmgreen and J. D. Ray burn (1987) "Social and Psychological origins of media use : a life style analysis". Journal of Broadcasting and Electronic Media 31 : 255-278.
8. Duraisamy, V (1977) "Psycho-linguistic analysis of farm broadcast" unpublished *M. Sc.* thesis submitted to TNAU, Coimbatore.
9. Gareet, J. (1969) "Statistics in Psychology and education" Bombay : Vakils, Feffer and Simons Pvt. Ltd.
10. Gumpert, G. and C. Ribert (1986) "Inter media-inter personal communication in media world, New York : Oxford University Press.

11. Joseph, D.A. (1978) "Communicology : an introduction to the study of communication." New York : Harper and Row Publication.

12. Klapper, J. (1960) "The effects of mass communications".

13. Glenose, Illinois : Free Press. Krishnakumar, K. N. (1982) "Effectiveness of farm journal articles," unpublished M.Sc. thesis submitted to TNAU, Coimbatore.

14. Levy, M.R. (1983) "Conceptualizing and measuring aspects of audience activity" journalism Quarterly 60 : 109-114.

15. Lin, N. (1977) "Communication effects : review and commentary in B Ruben (ed.) Communication Year Book 1. New Baunswick, Nj : transaction.

16. Manoharan, M. 1984 "Relative effectiveness of print media," unpublished Ph.D. thesis to TNAU, Coimbatore.

17. Muthazhagan, P. (1990) "Content analysis and readers perception on Valarum Velanmai and Seithi Madal Publications." Unpublished M.Sc. thesis submitted to TNAU-, Coimbatore.

18. Me Combins M. E. and D. Shaw (1972) "The agenda setting function of mass media". Public opinion Quarterly 44 : 545-548.

19. Murphy, R. D. (1977) "Mass Communication and human interaction", Boston : Houghton Miffilin Co. Meenakshi sundaram, K. S. (1990) "Television viewing behaviour of farm women". Unpublished M. Sc. thesis submitted to TNAU., Coimbatore.

20. Nagarajan, K. (1982) "An analysis or print media utilization by extension personnel". Unpublished M.Sc. thesis submitted to TNAU, Coimbatore.

21. Nataraju, M. S. (1991) "Effectiveness of farm magazines- a compinent analysis". Unpublished Ph. D. thesis to TNAU, Coimbatore,

22. Oliver, J. (1971) Impact of agricultural news disseminated by IADP through newspaper 'Dinamani' on the adoption of package of for paddy in Thanjavur district. Unpublished M. Sc. thesis submitted to TNAU, Coimbatore.

23. Rajendran, C. (1982) "Impact of broadcast through community radio sets". Unpublished M. Sc. thesis submitted to TNAU, Coimbatore.

24. Rathakrishnan, T. (1988) "Impact of agricultural telecast on farmers". Unpublished M. Sc. thesis submitted to TNAU, Coimbatore.

25. Seivaraj, G. (1990) "Effectiveness of Video teaching on farmers affective. Cognitive and psychomotor behaviour". Unpublished Ph.D. thesis submitted to TNAU, Coimbatore.

26. Sekar, V. (1982) "Farm broadcast behaviour of Extension personnel Unpublished M. Sc. thesis submitted to TNAU, Coimbatore.

27. Seshagiri, K. (1982) "Use of audio-learning in transfer of farm technology". Unpublished M.Sc. submitted to TNAU, Coimbatore,
28. Sridhar, V. R. (1983) "Farm telecast : An analysis of the impact and viewers preferences". Unpublished M. Sc. thesis submitted to TNAU, Coimbatore.
29. Tipton, L, R. O. Haney and J. R. Baschcart (1975) "Media agenda setting in city and state election campaigns". Journalism Quarterly 52 : 15-22.

Chapter **20**

Television Viewers Activity Behaviour towards Krishi Darshan Programme

Arun Kumar Singh, A. K. Singh, O.P. Mishra

In an agriculture dominated India, the TV has more responsibilities to discharge. A break through in the field of agriculture is impossible without an effective communication support to disseminate the necessary information like research findings. Speedy dissemination of the information to the farming community effectively is essential for bridging the gap between the agricultural scientists and the farmers. Television with its unique potential has contributed much through the "Krishi Darhsan Programme", which was introduced on January 26, 1967. TV has been acclaimed as a powerful audio-visual media today than any other media as it communicates effectively the information and technological know how to the farmers (Patil et al., 2000).

Although television has reached over 85.4 per cent of the Indian population and the country has more than 2 lakhs community TV sets in rural areas (Dwarakanath, 2002) by and large, the success of developmental programme like Krishi Darshan Programme depends on the audience and their activity behaviour. It is therefore, of interest to any development communicator to know what happens to the farm telecast information at the farmer's end. The farmers as audience of the Krishi Darshan Programme (KDP) are expected to receive the message with intention, purpose and utilize it effectively in their agricultural enterprises.

The active participation of the audience according to Tipton et. al. (1975) was critical factor for the assessment of the impact of mass media. Because audiences are variably active in their approach to media, audience activity is an important aspect of media uses and gratifications (Blumler, 1979 and Levy and Windahl, 1985).

Research Methodology

This audience research was carried out in the newly formed state 'Uttaranchal'. The district of Udham Singh Nagar and Rudrapur block was selected purposively for the study. Out of ninety villages of the selected block, two villages were selected randomly. Both the villages had a good number of television sets. All the TV owners belonging to the two selected villages were taken as respondents. The study was conducted through personal interviews with the help of well-structured pre-tested interview schedules developed for the study.

The audience activity postulates a voluntary and selective orientation by audiences towards the communication process. It suggests that the media use is motivated by needs and goals that are defined by audience members themselves and that participation in the communication process may facilitate, limit or otherwise influence the gratifications and effects associated with exposure (Lin, 1977). It is best conceptualized as a variable construct, with audiences exhibiting varying kinds and degrees of activity.

Based on the typology of audience activity suggested by levy (1983) three types of audience activity viz., pre-exposure selectivity, exposure involvement and post-exposure utility were considered as the constituents of the audience activity. The individual make goal-oriented choices about a given set of communication in the pre-exposure selectivity phase. The second type of audience activity focuses on the degree of psychological involvement audiences exhibit during exposure and related to individuals and how they attempted to provide meanings with the message. **Levy** (1983) used two measures of involvement during exposure-attention and para-social interaction. The third type of audience activity, post-exposure utility refers to the activities of the individuals such as discussions after telecasts, adoption in their own fields i.e. integrating the information gained from the communication settings into psychological or social behaviour.

The statements developed by Krishna Kumar (1990) to assess the audience activity of mass media audience was used with suitable modifications to measure the audience activity behaviour of the Krishi Darshan Programme viewers. The final score of the audience activity was obtained by summing up the scores of the pre-exposure selectivity, exposure involvement and post exposure utility. Based on the scores, the KDP viewers were categorized into three groups as low, medium and high in terms of their audience activity.

Findings and Discussion

Pre-Exposure Selectivity

The first activity type links "selectivity" to the pre-exposure phase of the communication sequence. The individual makes choice about the given set of communication based on their learned experience (Levy, 1983).

Table-1. Distribution of viewers of KDP in terms of their audience activity

Activity type	Low	Medium	High
Pre-exposure selectivity	32 (25.36)	77 (61.12)	17 (13.49)
Attention	19 (15.08)	83 (68.87)	24 (19.05)
Para-social interaction	23 (18.26)	84 (66.67)	19 (15.07)
Post-exposure utility	23 (18.26)	92 (73.01)	11 (8.73)

Note : Figures in the parenthesis indicates the respective percentage.

The pre-exposure selectivity of the majority of the viewers (61.12%) showed medium level (Table 1). The pre-exposure selectivity was measured in terms of the functions of viewers of the Krishi Darshan Programme as given in Table 2. The functions included both the positive and negative sentences.

Table-2: Distribution of the respondents according to their responses with respect to pre-exposure selectivity

S. No.	Statements	Always	Sometimes	Never
1.	I usually watch the clock so I would not miss the KDP	34 (26.98)	49 (38.89)	43 (34.12)
2.	I put on TV well in advance so as not to miss the KDP	27 (21.42)	52 (41.26)	48 (38.09)
3.	I use to go through the programme chart to know the subjects to be dealt in KDP	41 (32.53)	49 (38.89)	36 (28.57)
4.	I use to view the KDP accidentally while I view other programme	41 (32.53)	67 (53.17)	18 (14.28)
5.	I keep everything away to view the KDP if the subject of my interest is being dealt with	30 (23.80)	56 (44.45)	40 (31.75)
6.	I use to note down the KDP schedule in may diary so as not to miss the useful programme	10 (7.93)	31 (24.60)	85 (67.46)
7.	I ask my family members to remind me at the time of KDP	10 (7.93)	42 (33.34)	74 (58.73)

Note : Figures in parenthesis indicates percentage in their respective category.

This selective exposure behaviour was said to be based on the learned perceptions of the media and prior experience with them, the individual had come to expect certain level of gratification (Levy and Windahl, 1985).

The moderate level of pre-exposure selectivity infers that they expect a moderate satisfaction from the Krishi Darshan Programme, and its contents.

Exposure involvement

Involvement is seen as a dimension of audience activity during message reception that shows the intellectual and emotional participation in message content. It isa state of activation that is being reflected in attentiveness and para-social interaction.

(a) Attention

The attention paid by the viewers of KDP was best reflected by their attentive and undistracted viewing rather than notes taking or making salient points while watching KDP. Majority of the viewers (51.58%) gave undivided attention always while watching the KDP (Table 3). This was in line with the findings of Laharia and Joshi (1992) and Pochaiah and Reddy (1998).

Table-3 : Distribution of the respondents according to their responses for statements indicating attention (N=126)

S. No.	Statements	Always	Sometimes	Never
1.	I use to view the KDP from the beginning to end	52 (41.26)	64 (50.79)	10 (7.93)
2.	I give undivided attention while I view the important KDP	65 (51.38)	34 (26.98)	27 (21.42)
3.	I take notes while viewing the KDP	13 (10.31)	35 (27.78)	78 (61.90)
4.	I use to talk with others while viewing KDP	39 (30.95)	69 (54.76)	18 (14.28)
5.	I use to do other works while viewing KDP	47 (37.30)	63 (50.00)	16 (12.69)

Note : Figures in parenthesis indicates percentage in their respective category.

The overall score of the activities related to 'attention' resulted in the medium level of attention by the majority of the viewers (Table 1).

(b) Para-social interaction

Levy (1983) found that para-socially involved viewers were actively involved in making the meaning of the media messages.

Table-4 : Distribution of the respondents according to their responses for statements indicating para-social interaction

S. No.	Statements	Always	Sometimes	Never
1.	KDP makes me feel at a very familiar situation	71 (56.34)	33 (26.19)	22 (17.46)
2.	I feel as if some one in my family talking to me when I view the KDP	44 (34.92)	33 (26.19)	49 (38.89)
3.	The programme telecaster have become almost like good friend or people I know	48 (38.09)	29 (23.01)	49 (38.89)
4.	I feel as if the messages of KDP are almost meant for me	37 (29.36)	52 (41.26)	37 (29.36)
5.	KDP appear to be more instructive rather than persuasive	60 (47.61)	54 (42.85)	12 (9.52)
6.	I am not able to identify myself with the KDP telecasters	60 (47.61)	51 (40.47)	15 (11.90)
7.	KDP persuades me to interact with the telecaster	18 (14.28)	44 (34.92)	64 (50.79)

Note : Figures in parenthesis indicates percentage in their respective category.

The respondents mostly felt the KDP at a very familiar situation and the programme telecaster have become almost like good friend or the people they know (Table 4).

Moreover, the Table 1 highlighted that most of the respondents (66.67%) exhibited the medium level of para-social interaction.

Post-exposure utility

Table-5 : Distribution of the respondents according to their responses for statements indicating post-exposure utility

S. No.	Statements	Always	Sometimes	Never
1.	I discuss with others what I have viewed in the KDP	44 (34.92)	57 (45.23)	25 (19.84)
2.	I communicate or recommend the messages viewed through KDP to other famrers	32 (25.39)	70 (55.56)	24 (19.04)
3.	I use to think about and evaluate the messages viewed in the KDP	48 (38.09)	52 (41.26)	26 (20.63)

S. No.	Statements	Always	Sometimes	Never
4.	The practices advocated in the KDP are not applicable to my field conditions	51 (40.47)	54 (42.85)	21 (16.67)
5.	I use to note-down useful points, which I remember after the KDP	14 (11.12)	36 (28.57)	76 (60.31)
6.	I try to adopt the practices advocated in the KDP	39 (30.95)	46 (36.50)	41 (32.53)
7.	I use to correspond with the KDP telecasters whenever I need clarification or more details necessary for adoption	6 (4.76)	33 (26.19)	87 (69.04)

Note : Figures in parenthesis indicates percentage in their respective category.

The extent to which an individual utilize the messages acquired in the communication setting during his post-exposure social life or psychological life determines the effectiveness of that particular communication process. The responses regarding the activities related with the post-exposure utility like discussion, adoption, feedback is given in Table 5.

The discussion, recommendations to other farmers, adoption of the practice telecast in their own fields were found to be at moderate level. Also, from Table 1 it is evident that the majority of the respondents (73.01%), exhibited medium level of post-exposure utility.

Audience activity

The receiver-oriented concept of audience activity is fundamental to the study of mass communication effects. Hence, the study focuses on the concept and the total audience activity score was obtained by adding the scores of the individual activity types viz. Pre-exposure selectivity, Exposure involvement (Attention and Para-social interaction) and Post-exposure utility.

Table-6. Distribution of the respondents according to their level of audience activity

S. No.	Audience activity	Number of respondents	Percentage
1.	Low	20	15.87
2.	Medium	86	68.26
3.	High	20	15.87

A vivid inference could be evinced from the Table 6, that two-third of the respondents (68.26%) had medium level of audience activity. It was clear from the individual activity types itself, as all of them exhibited only the medium level of activity. Both the high and low level of audience activity had a share of 15.87% of the respondents respectively.

Conclusion

The study was concerned with the activity behaviour of the farming community on viewing KDP. On the basis of the findings and dominant trends it can be concluded that the audience activity of the KDP indicates the effectiveness of the programme and it was found to be at medium level for the majority of the respondents. It had scope for the improvement by producing more interesting KDP having location specific and need-based approach. Proper co-ordination between the extension agencies and rural broadcasters will make KDP more purposive and meaningful.

References

1. Blumer, J.G. (1979). The role of theory in uses and gratifications studies. Communication Research. 6:9-36. Dwarakanth, H.D. (2002). Mass media: Crucial role in promoting rural development.
2. Kurukshetra. 50 (6):4-10. Krishna Kumar, K.N. (1990). Farm Information Agenda Setting-Comparative Media Analysis. Unpublished Ph.D. Thesis, Tamil Nadu Agricultural University, Coimbatore.
3. Laharia, S.N. and Joshi, N.N. (1992). Farm telecast viewing behaviour of farmers. Indian Journal of Extension Education .28 (1&2): 15-22. Levy, M.R. and Windahl, S. (1985). The concept of audience activity. Media Gratification Research.
4. Levy, M.R.(1983). Conceptualizing and measuring aspects of audience activity. Journalism Quarterly.60:109-114.
5. Lin, N. (1977). Communication effects: Review and commentary. In Ruben, B.(ed.) Communication Year Book . 1. New Brunswick, NJ: Transaction.
6. Patil, C.B.; Jondhale, S.G. and Bhosle, P.B. (2000). Correlation of perceived effectiveness of Farm Telecast. Maharashtra Journal of Extension Education. 19:37-40.
7. Pochaiah, M. and Reddy, P.K. (1998). Listening and viewing behaviour of farmers towards agricultural programme. Maharasthra Journal of Extension Education. 7:307-311.
8. Tipton, L.; Haney, R.O. and Baschcart, J.R. (1975). Media agenda setting in city and state election campaigns. Journalism Quarterly. 52: 15-22.

Chapter 21

Role of Mass Media of Communication in Disseminating Farm Information- A Comparative Analysis

B.M. Thombre and M.G. Bhilegaonkar

Since two decades there has been substantial growth in the knowledge of agriculture specially improved methods of cultivation. The benefits of newly emerging body of knowledge can, however be realized only if it is shared with farmers, the ultimate users. Efforts in the direction to arrange for supply of inputs like seed, fertilizer, power and water for agriculture is not enough but there is need to provide information input also to increase present level of agriculture production. Information play an immense role in our society and its large scale dissemination through highly preferred and credible communication media constitutes one of the most striking development of our time. Information has become an integral part of our daily life. The objective of communication of the agricultural information to farmers is not only to create awareness among them but also to impart new ideas that change their behaviour and mode of farming.

It is well established fact that through mass media it is possible to reach large number of people to raise their interest in farming (Verma and Yadhav, 1974). Progress and change in farming community can only emerge by timely and rapid communication of the farm information. Various mass media are in use out of those prominent are the Radio, Farm literature, Television, Films and Agricultural exhibitions. These media of communication possess the potentiality to convey scientific information on farm practices quickly and promptly with high fidelity. No doubt the fast growing technologies requires a sound and well planned system of communication. Mass media play significant role in dissemination of agricultural Technologies. This has been revealed in past studies like Kishore (1968); Sandhu and Singh (1972) and Dey (1981).

The development in mass communications and its choice will very much depend on the available resources and types of audience. Therefore, the present study was an attempt to investigate in to the profile of mass media users, the extent of use of the mass media, its credibility pattern, and the relationship existing between extent of use of mass media and characteristics of users. The study will also be attending to identify the problems in use of mass media and suggestion to improve its use in order to make these information sources effective.

Following are the objectives of the study.

1. To study the Profile of users of Mass media in farming community.
2. To know the extent of use of Mass media and its preference by the farmers.
3. To study the comparative credibility of selected mass media.
4. To find out existence of relationship, if any, between the extent of use and the characteristics of the users.
5. To identify the problems in use of mass media sources and suggestions to improve its use.

Material and Methods

The present study was conducted in college of Agriculture, Parbhani (Maharashtra) Extension Block. Ten villages were randomly selected from two panchayat samiti blocks namely Parbhani and Basmath. Stratified random sampling method was used to draw a sample of 150 respondents from these selected villages. The data were collected through personal interview with the help of a structured schedule.

The profile of the mass media users were studied with the help of open and multiple choice questions. The respondents were asked to give their responses on three point scale like always, some times and rarely about the use of mass media for getting agricultural information. The score of 3, 2 and 1 respectively was assigned. Total score thus obtained were categorised. The credibility pattern of mass media was measured on three point continuum namely more credible,

credible and less credible for each media and score of 3, 2 and 1 was assigned respectively. Total scores thus obtained were categorised. The data thus obtained was then tabulated and analysed by using frequencies and percentages. The correlation co-efficient were computed to find out the relationship between mass media users and their characteristics.

Results and Discussion

Profile of Mass Media users

It was observed from table 1 that 52 percent of the mass media users were between 31 to 45 years of age, while 1/3rd of them (30.67 percent) were having age above 45 years. As regards education it was noticed that 25.33 per cent of the users did not have formal education. Only 21 .33 per cent users had education up to high school and college level. Land possession pattern of the mass media users indicated that 63.33 per cent were having land above 4 ha. The data of annual income of the users proved that 42 per cent had annual income between Rs. 2,001 to 6,000. About twenty five per cent media users annual income ranged between Rs. 6,001 to 10,000. One fifth of total users' income was more than Rs.10,000. As regards type of family, it is observed that 52 per cent mass media users family was nuclear and of remaining 48 per cent had joint family system.

Information regarding social participation level indicated that 2/3rd of the mass media users had low social participation. Data regarding overall modernity levels of the mass media users indicated that half of the users were in medium level of overall modernity, while 28 per cent had high level of overall modernity.

Extent of Mass Media Use

The extent of use of some of the important mass media by the respondents for getting agricultural information indicated (Table-2) that radio was always used by 44.67 percent of the respondents. It was followed by 34.67 per cent of them using farm literature frequently. Similarly, it was also observed that 49.33, 40.67, 32 and 20.67 per cent of the respondent were using radio, farm literature, films and agricultural exhibitions some times as the source of farm information respectively. It was expressed by 66.33 per cent users that they rarely visited agricultural exhibitions for getting information.

Table-1. Distribution of Mass Media users according to socio-personal and psychological characteristics.

S. No.	Characteristics of Mass media users	Frequency	Percentage
	Age (years) :		
1.	Up to 30	26	17.33
2.	31 to 45	78	52.00
3.	Above 45	46	32.67
	Education :		
1.	Illiterate	38	25.33
2.	Primary school	58	38.67
3.	Secondary school	22	14.67
4.	High school and above	32	21.33
	Land holding :		
1.	Small (up to 2 ha)	27	18.00
2.	Medium (2.1 to 4 ha)	28	18.67
3.	Big (above 4 ha)	95	63.33
	Annual income (Rs.) :		
1.	Up to 2,000	19	12.67
2.	2,001 to 6,000	63	42.00
3.	6,001 to 10,000	37	24.66
4.	Above 10,000	31	20.67
	Type of family :		
1.	Nuclear family	78	52.0
2.	Joint family	72	48.00
	Social participation level :		
1.	High (above 3)	11	7.33
2.	Medium (2 to 3)	45	30.00
3.	Low (0 to 1)	94	62.67
	Over all modernity level :		
1.	High (above 25)	42	28.00
2.	Medium (21 to 25)	75	50.00
3.	Low (up to 20)	33	22.00

Table-2. The extent of use of the select Mass Media by the farmers.

S. No.	Mass media	Frequency and % of use			Total score obtained	Rank order
		Always	Some time	Rarely		
1.	Radio	67 (44.67)	74 (49.33)	-	540	I
2.	Farm literature	52 (34.67)	65 (40.67)	14 (9.33)	362	II
3.	Films	-	48 (32.00)	69 (46.00)	319	III
4.	Agricultural exhibitions	-	31 (20.67)	95 (63.33)	278	IV

Figures in the parenthesis indicate percentage.

The findings of the study indicated that radio was most used (preferred) media, followed by medias[1] like farm literature, agricultural films and agricultural exhibitions in order of its preference in getting agricultural information. These findings are in corroboration with results of Kishore (1968), wherein it was reported that radio was proved to be more effective means of communication. Similarly studies of Moulik and Lokhande (1970) and Sadamate and Sinha (1976) have also confirmed these findings by quoting that radio played very important role in imparting the farm information.

Credibility pattern of Mass Media

The relative credibility for four important mass media as per used by the respondents indicated (Table-3) that radio (60.67 per cent) was more credible source. This was followed by farm literature (22.67 per cent) and films (16.67 per cent) in order of its ranks in per using its credibility.

Table-3. The comparative credibility for selected Mass Media.

S. No.	Mass media	More credible	Credible	Less credible	Total score	Credibility rank
1.	Radio	91 (60.67)	42 (28.00)	17 (11.53)	374	I
2.	Farm literature	34 (22.67)	61 (40.67)	55 (36.67)	279	II
3.	Films	25 (16.67)	46 (30.67)	79 (46.00)	246	IV
4.	Agricultural exhibition	27 (18.00)	49 (32.67)	74 (49.33)	253	III

Figures in the parenthesis indicate percentage.

These findings are in line with the findings of the Sonoria and Singh (1970) and Singh and Singh (1971). Where it was reported that radio was more credible source for the farmers. Ascribing radio as most credible source by the farmers might be because of radio broadcast by the agricultural scientists. More over low literacy percentage in farming community may act as a barrier to use printed media as information source.

Relationship between use of mass media and Profile of users

The data regarding existence of relationship between the extent of use of mass media and the Socio-personal and psychological characteristics of the farmers indicated (Table-4) that age of the farmers had negative significant relationship with the extent of use of mass media. While education, annual income, size of Land holding, Social participations, Level of overall modernity and credibility has positive influence on the use of mass media. That means farmers with better socio-economic status were modern in outlook which might have induced them to use of mass media. These findings had a support from Kishore (1968) and Dey (1981).

Table-4. Relationship between extent of use of Mass Media with characteristics of users.

S. No.	Characteristics	Correlation coefficient
1.	Age	0.315**
2.	Education	0.288**
3.	Annual income	0.425**
4.	Land holding	0.353**
5.	Social participation level	0.331**
6.	Overall modernity level	0.436**
7.	Credibility	0.373**

** Significant at 0.05 level of Probability.

Problems in Mass Media Use

In order to know the problems of mass media users, they were exposed to a structured Schedule. The data indicated (Table-5) that 70.67 per cent of the mass media users expressed the problems like broadcasted radio programme, was not repeated and hence they were not easily understood.

It was observed in case of farm literature use, that 50.67 per cent of users expressed the non-availability of farm literature in time and in village. This was followed by illiteracy, language used in farm literature was difficult to understand and printing was not readable.

S. No.	Problems	Frequency	Percentage
A	Radio :		
1.	High prices of radio	89	59.33
2.	Language used in broadcast in difficult to understand	34	22.67
3.	Message once broadcast is not repeated	106	70.67
4.	Timing of agricultural programme is not suitable	21	14.00
B	Farm literature		
1.	Illiteracy	38	25.33
2.	Language used was difficult to understand	23	15.33
3.	Non-availability of farm literature in time and village	76	50.67
4.	Printing is not readable	19	12.67
C	Films		
1.	Film show are not organized frequently	116	77.33
2.	Films on family welfare were only	97	64.67
3.	Films does not have rural back-ground	126	84.00
4.	Films are difficult to understand	9	6.00
D	Agricultural exhibition		
1.	Agriculture exhibition are not organized in village or nearby places	21	14.00
2.	Required information is not available always	48	32.00
3.	Exhibition are not arranged frequently	141	94.00
4.	Proper samples were not depicted	11	7.33

In the use of agricultural films, it was expressed by 84 per cent of users that the films do not have rural background hence do not attract rural people. This was followed by 77.53 and 64.67 per cent of media users expressing the problems like film shows were not organised frequently and films on family welfare were only screened rather than of agriculture respectively. It was very interesting to note that 94 per cent mass media users expressed that exhibitions were not organised frequently in village or nearby place.

Suggestions of Mass Media Users

The Mass Media users were also asked to offer their suggestions in order to overcome the problems in its use. It was observed (Table - 6) that 61 per cent users suggested that agricultural exhibitions had to be organised frequently in the villages.

Table-6 - Suggestions given by Mass Media Users

S. No.	Suggestions	Frequency	Percentage
1.	Increase the duration and repetition of farm broadcast programme	52	34.67
2.	Publications should be made available in the time and in village	38	25.33
3.	Film shows showed be organized frequently in the village	48	32.00
4.	Films with rural background on farm information should be screened	81	52.00
5.	Agricultural exhibitions should be organized frequently	92	65.33

Another important suggestion made by 54 percent users was to screen the films having rural background for easy understanding. Some other suggestions like increase the duration and repetition of farm broadcast (34.67 per cent), frequent organization of film shows on agricultural topics (32 per cent) and farm publications to be made available to the farmers in time and in villages were also made. Similar finding was also reported by Pande and Roy (1977).

Implications :

In the light of the findings emerged, it can be suggested that radio as mass media need to be used frequently, by increasing the present duration and repetition of agricultural programmes through the regular broadcast. It was also observed that farm literature could enthuse some literate farmers therefore, its quality, content and availability need to be improved, so that, its use at village level can be increased. The rural institutions like Gram Panchayat, rural schools and co-operative societies be enrolled of mailing list of print media distributed. This will help to increase the use of print media by the members of those institutions. This will also help to develop secondary readership in village where illiteracy percentage is more. The film shows and agricultural exhibitions need to be organised frequently in rural community.

REFERENCES

1. Dey, P. K., Channels preference for agricultural message, Indian Joi. Exm. Eden. 17 (3 and 4) : 17-21, 1981.
2. Kishore, D., A study of effectiveness of radio as *a* mass communication medium in dissemination of agricultural information, Ph.D. Thesis (Unpub.) I.A.R.I., New Delhi, 1968.
3. Moulik, T. K., and M. R. Lokhande, Radio in village, Kuruk-shetra, 18 (5): 15-16, 1970.
4. Pande, S. N. and N. K. Roy., Relative effectiveness of three modes of presentation of farm broadcast, Indian J. of Exm. Eden, 13(3 & 4) : 45-49, 1977.
5. Sadamate, V. V. and B. P. Sinha, Dissemination of farm information through Television, Radio and Extension Agency - A comparative study, Indian J. of Extn. Eden. 12 (1 and 2) :1 5, 1976
6. Sandhu, fl-S. and K.N.Singh, Listening habits of Radio owning farmers in respect of farm programmes, Indian J.of Extn. Eden, 8 (1 and 2); 30-56, 1972.
7. Singh, Daulat and M.R-Singh, Farmers perception of different credibility pattern of information, Indian J.of Extn.Eden. 8(3 and 4): 78, 1971.
8. Sonoiia, Y.C., and K.N.Singh, Communication sources and their credibility, Indian J.of Extn. Ed en. 13 (3 and 4): 22-23, 197Q.
9. Verm a, P.5. and l.P.Yadhav., Agricultural communication and its prospectus in India, Rural India, 38 C> and 6): 74-76, 1974.

Chapter 22

Farm Communication through Television: An Overview

A. Ratnesh Babu and A. U. Gadewar

Modern electronic media viz., colour television, telecast via satellite. Community Television System (CIS], Video Cassette Recorder (VCR), Video Cassette Player (VCP), Cable TV etc., have completely revolutionized the communication system in India. Of these, the television and video have a profound impact on people's socio-economic, political and cultural life. Television provides an immense learning experience. In the words of Schramm (1953), "People would certainly learn from television and learning from television would not remain confined to facts only, but it extends to development of skills, critical thinking and problem solving".

Television in India

The origin of television dates back to 1936, during which the first regular television transmission was started by B.B.C. The public of the United States had its first view of television at World Fair in New York as back as 1939. In India, television was first inaugurated on September 15, 1959 on and experimental basis in the form of UNESCO's pilot project in Delhi. Twenty sets were installed at selective community centers in an around Delhi and the duration of the programme was for one hour twice-a-week. The regular television service was introduced on the independence day of 1965, with one hour duration daily.

Krishi Darshan the first rural programme was started on January 26, 1967. The basic objectives of this programme are to create awareness among the rural viewers with the latest technical and scientific know-how with regard to crop cultivation practices; fertilizers; soil testing; dairying; animal husbandry; weather forecasts, etc., and to enlighten rural audience about the importance of education, personal hygiene, health and family welfare, and finally to provide entertainment. Today, farm telecasts are transmitted from all the metropolitan cities, post SITE, INSAT, centres and low power transmitters of *Doordarshan*.

The Satellite Instructional Television Experiment (SITE), one of the most exciting projects, was launched on August 1, 1975. It was a one year project covering 2400 villages spread over six states of India with four linguistic groups. Several of the research studies indicated the success of SITE for having contributed substantially to the gain in Knowledge and a positive change in the attitude of rural viewers.

Doordarshan introduced its first own multipurpose satellite known as INSAT in 1982. From the Independence Day of the same year, Doordarshan started its national programmes, related simultaneously from all its centres. The INSAT telecasts are primarily aimed at providing programmes of local relevance and education for rural and tribal people living in remote and difficult Farm Communication areas of the country. Most of the INSAT programmes are locally produced in difference formats. These include folk dance, plays, skits, puppetry, development reports from the field etc.

Satellite telecasts through the cable TV system started with the coverage of the Gulf War in early 1991 by CNN. In October 1991, STAR TV started showing the BBC World Service for 24 hours a day in India. An IIMC study (Dayal, 1992) reveals that the viewership of these foreign programmes beamed by STAR TV is substantial.

Profile of Rural TV Viewers

Studies on characteristics of TV viewing farmers indicated that most of the viewers belong to the age group of 31-50 years (Dey, 1968; Sadamate, 1975; and Sahi, 1977). More than two-third of the viewers found to have undergone formal schooling up to middle standard and above (Singh, 1971; and Sadamate, 1975). Contrary to it, Sahi (1977) observed that a major section of the rural TV viewers were illiterate and could not read or write.

Majority of the rural TV viewers (72%) belonged to lower income group (Singh, 1971; and Sahi, 1977). Nearly 67% of the viewers belonged to large family size and more than 75% of the viewers were heads as well as decision makers of the family (Sahi. 1977; and Jha, 1978).

Seventy percent of rural TV viewers seem to be cosmopolite than the non-viewers (56%). Nearly 64% viewers had high media exposure as against 46% for non-viewers (Singh, 1971; and Sadamate, 1975).

Several studies reported that the viewers had more favourable attitude towards television than the non-viewers. An overwhelming majority i.e. 92% had favourable attitude towards television as against only 34% of non-viewers (Sadamate, 1975). Similar observations were made by Singh (1971) and Singh (177). In Televised area, the attitude of farm women were more favourable as compared to the farm women of non-televised area with regard to the grain storage and nutritional practices (Dubey and Swarnakar, 1985).

Viewing Behaviour

The average period of viewing television by the heads of households was found to be 48 minutes per day and the corresponding figure for their wives was 65 minutes per day (Jha, 1978). However, Mehra et al. (1985) reported two-third of their respondents viewed television on an average for 1-2 hours per day and the remainder for 2-3 hours. They further stated that majority of the farmers were spending more time on discussing what has been seen on television when compared to any other mass media.

Education was not a limiting factor for consumption of information through farm telecast (Sadamate and Sinha, 1977). In contrast, Singh (1971) and Singh (1977) observed that education has significantly contributed to the viewing behaviour of the rural audience. However, Kumar et al. (1991) through a meta analysis found that only 28.5% of the television studies reported positive correlation between TV viewing behaviour and education while 70.5% of the studies ruled out such possibility.

Farmers need for achievement and change proneness were found to be the most significant factors contributing towards farm televiewing behaviour (Sinha and Mehta, 1971 and Singh, 1977). Attitude of the farmers and media exposure were found to be closely and significantly associated with farm televiewing behaviour of the rural audience (Singh, 1971; Singh, 1977 and Sadamate and Sinha, 1977). However, age and distance have negative significant relationship with the TV viewing behaviour (Kumaretal., 1991).

With regard to frequency of viewing television, 31% of farmers viewed *Krishi Darshan Programme* almost regularly but there were 21% who had viewed the programme only occasionally while 22% were not much regular. Close to the percentage of regular viewers were the farmers (26%) who had never viewed the programme (Mehra et al., 1985). Singh (1992) reported that majority of the farmers have possessed knowledge about frequency of the *Krishi Darshan* Programme. Cherian and Chandra (1989) observed that the extent of watching the television programmes was more in case of men than women.

Another important areas could be of finding out the reason for non-viewing of the *Krishi Darshan Programme.* Sadamate (1975) reported as many as eight reasons i.e., small size of land holding (28%), village factions (20%), farming being a subsidiary occupation (14%), physical incapability (10%), unfavourable perception (10%), lack of time (8%), poor comprehension (6%), and hostility due to non-availability of inputs (4%).

Low attendance at *Krishi Darshan Programme* was due to the conspicuous absence of entertainment appeal (Singh, 1971). In addition, there was a host of other reasons such as viewers were critical about the technology being shown on TV which they felt were not always within the available resources, low level of understanding of an average farmer; lack of knowledge about the scheduled days, timing, type of *Krishi Darshan Programme* to be telecast, absence of extension personnel to explain the programme and lack of general ability of some farmers to derive meaning out of visual images telecast in quick succession. Surprisingly, several studies reported that the rural TV viewers were not aware of the days and timing of telecast of farm programmes (Singh, 1971; Singh et al. 1975; Chouhan and Sinha, 1976; and Sangha et al., 1979).

With regard to the negative views held by the respondents about television viewing, Mehrotra and Aujla (1991) reported that 81% of their respondents expressed 'television disturbs studies of children some times and 'give them unwanted exposure'. The other negative views expressed by the respondents were as follows: 'late night sleep' (76%), 'unplanned meals time' (67%), 'reduction in informal visits of friends and relatives' (76%), and 'less attention to important productive activities' (59%).

Television advertisements are found to be not compatible to socio-cultural set-up of the rural people. By and large families lacked orientation in the right perspective towards TV advertisements and as such held unfavourable attitude about them (Chaturvedi, 1992).

Programme Production

The various aspects of farm programmes production such as mode of presentation, time of telecast, duration, frequency and content of farm telecast were studied by several research scholars.

'Specialists in conversation with farmers' was found to be the best mode of presentation for farm telecasts (Sekhon, 1970; Sadamate and Sinha, 1977; and Mhurkar, 1985).

Interestingly, farm women have a different choice on the timings of *Krishi Darshan Programme.* As high as 88% farm women preferred the programme to be telecast from 8.00 PM onwards. Only 12% of them were in favour of keeping the present timing (Kaur, 1970).

Over 60% of the farm televiewers expressed that the duration of *Krishi Darshan Programme* to be increased to one hour (Sekhon, 1970. Similar observations were made by Mishra (1967); and Cnounan and Sinha (1976).

With regard to frequency of *Krishi Darshan Programme,* majority of the farmers expressed their views that the programme should be telecast daily (Sekhon, 1970). Probably keeping in view such findings and concern, the *Doordarshan* is now telecasting *Krishi Darshan Programme* daily. In terms of repetition of message for reinforcement, farmers felt that the same programme should be telecast with an interval of 15 days (Sharma and Day, 1970).

Timelines of the message is one of the important aspects of programme production. Kauar (1973) explicitly observed that the programmes in agriculture will have an immediate effect or at least catch the imagination of the viewers when they are telecast at the time when farming community needs them most during various crop stages. Several of the studies have highlighted that the programme producers were not fully conversant with the exact information need of the viewers (Singh and Prasad, 1975 and Singh, 1985).

Dairy and animal husbandry programmes were found to be as low as eight percent of the total telecast of the *Krishi Darshan Programme* in a year (Singh, 1985). Chauhan and Sinha (1976) pointed out that there was less information on plant protection aspects in *Krishi Darshan Programme.* Nearly 90% of the farm women felt difficulty in following the language being used. Technical words such as use of metric system of weights and measures are not followed by them. As many as 87% of the viewers complained the use of English words in dialogues (Kaur, 1970). Similar findings were reported by Chauhan and Sinha (1976).

Some meaningful entertainment through *Krishi Darshan Programme* for better reception of the farm message was suggested by Singh, 1971; Singh and Sharma, 1973; and Singh, 1992.

In order to overcome some of the above shortcomings, Singh and Prasad (1975) suggested a close collaboration among the Television Centres, State Agricultural University, Indian Council of Agricultural Research and State Department of Agriculture in **farm** programmes production By virtue of their professional competence, the quality of farm programmes can be ensured.

Development of human power for TV in areas of programme production and communication research needs far more attention, priority and resources (Karnik, 1989). Subbramanyan (1991) suggests that *Doordarshan* must strive hard to telecast more TV programmes of educational nature. Kashyap (1991) emphasizes the need to improve the quality of *Doordarshan* programmes and suggests that its unique identity will have to be evolved and for that it should have a suitable mechanism.

Effectiveness of Television

Television is the most effective medium known today for mass education in the process of national development. Research studies on gain in knowledge, credibility, TV in comparison with other media, and its influence on adoption of technology **revealed** interesting information. They **are as** follows:

Gain in Knowledge

Evaluating the impact of television on farmers in relation to gain in knowledge and retention of gained knowledge, Mishra (1967) reported that there was significant increase in knowledge from 10.12 to 28.15 percent among the farmers exposed to agricultural telecasts. The average gain in knowledge was reported to be 20.27%. Retention of gained knowledge after 15 and 30 days was also significant i.e. 79.04 and 67.44 percent respectively.

Further, Sinha (1970) observed a high level of gain in knowledge through farm telecast. The farmers gained 24-90 percent of the information imparted through telecast with an average of 54.12%. They retained 72 to 100% of the knowledge gained after a lapse of 15 days with an average of 82%.

The farm women televiewers gained knowledge to the extent ranging from 43-73% of the information telecast through *Krishi Darshan* (Kaur, 1970). Similar findings were reported by Sekhon (1968), NCERT report (1969), Dubey and Swarnkar (1985) and Singh (1992).

On an average higher the education more was the gain in knowledge and its retention among rural television viewers (Dey, 1968; Sekhon, 1968; and Kaur, 1970). However, Chauhan and Sinha (1976) reported that the level of formal education was found non-significant related with gain in knowledge through farm telecast. But definitely education will help in choosing the appropriate information.

The socio-personal and economic characteristics of farm women were reported to be significantly associated with the extent of knowledge in televised area (Dubey and Swarnkar, 1985). Farmer's achievement motivation and change proneness were reported to have a significant relationship with their acquisition of information from a farm telecast (Sinha, 1970; and Singh, 1977).

Television helped men and women to gain significant amount of knowledge about green leafy vegetables, polio, vaccination and laproscopy (Cherian and Chandra, 1989). The post and pre-telecast knowledge scores were significant at 0.01 level among both men and women. They also found that the gain in knowledge was more among women than men.

Raghupathy and Talawar (1990) in their study on Mushroom cultivation practices found that the mean knowledge score of experimental group before viewing the telecast was 3.37 which had gone up to 20.17 after viewing. While in the control group, the mean score before was 3.57 and the gain in knowledge was marginal i.e. 4.37. The t' value revealed that there was significant difference between the two mean knowledge scores of experimental group while in the control group there was no significant difference.

Credibility

With regard to credibility of television as a source of information Sangha and Gupta (1985) observed that television occupies first place followed by radio, newspapers and government publications.

Mahurkar (1985) studied credibility of seven different sources of farm information and reported that television ranked first as the most credibility source. Similar observation was made by Singh and Sharma (1973).

Television In Comparison with Other Media

On an average television viewers and retained higher knowledge than the radio listeners in all the exposures (Dey, 1968). Telecasts relating to demonstration of new skills and methods can best be presented through television as compared to radio (Sharma and Dey, 1970).

Singh (1971) redicated that the television has a substitution effect over radio. With the introduction of nationwide television coverage it tends to isolate radio with the function of news and entertainment programmes for the audience. Strikingly, majority of the televiewers after owing television spends more time on it than any other mass media (Mehra et al., 1985). With regard to combination of media, Chauhan and Sinha (1976) suggested that the TV exposures, if supplemented with printed matter and followed by group discussion will have better impact on information acquisition and utilization by the farming community.

Adoption of Technology

Majority of the farmers with high need for achievement and change proneness showed a tendency towards the adoption of innovations telecast through *Krishi Darshan Programme.*

Whereas the farmers with low levels of need for achievement and change proneness showed a tendency towards rejection of the telecast innovations (Sinha and Sharma, 1972).

Television has significantly contributed for an increase in adoption of farm practices and helped particularly the non-progressive farmers to adopt more innovations (Sadangi and Rajguru, 1985). The farmers of television villages were far ahead of their counterparts of non-television villages with respect to adoption of high yielding varieties of wheat (Singh and Sharma, 1973). Similar observations were made by Sharma and Dey (1970); Choudhary (1985) and Singh (1992).

Information dissemination from viewers to non-viewers is considered as one of the important partners for measuring the effectiveness of farm telecasts. To this effect it was reported that a Farm Communication majority i.e. 76% of the viewers passed on information to non-viewers, thereby supporting the theory of two-step flow of information through mass media (Sinha, 1973).

Recent Advances

While explaining 'scenario cartography' in agricultural extension, Singh (1990) indicated that it is possible to image an information centre at Block level equipped with a computer, video cassette recorder and a monitor. The computer stores information related to each farmer of the block i.e. covering the information about his farm such as soil type, material and non-material resources, level of animal husbandry as well as the possible technological information relevant to the block's bio-physical and socio-economic conditions. The farmer asks for the plan. The information officer gets a print out of the farm plan. When the farmers reads the details of the farm plan, he expresses inability on certain skill parts. The information officer switches on VCR and the farmers observe the entire cultivation practices with emphasis on impact points on the monitor. He returns back home carrying with him his farm specific information printed on a sheet.

Video can be effectively used in the present era of rural development with its unique qualities of audio-visual recording, ease in handling and instant feedback, serves as an effective tool for rural extension work (Kishore Kumar, 1995). The author reports that video has been used to develop agricultural production skills of farmers and extension agents. Video cassette have been produced for various types of package of practices based on agro-climatic conditions. Practices which involve various steps and skill can be demonstrated very efficiently and repeatedly with the video. The Tamil Nadu Agricultural University, Coimbatore, made a venture in the production of video cassettes which narrate various practices and operations involved in the crop production and protection aspects. The communication centre of the T.N.A.U., lending video cassette to the farmers for their viewing. The other agricultural universities also must make an attempt in this direction.

Interactive video utilizes video and computer technologies to present information and to interact with the learner. Because of its capability to invite response and act on the answers which are made, the experience of learning from interactive video can be highly motivating and educationally very effective. Talawar (1994) suggested the use of interactive video in agricultural training system. As most of the agricultural training centres viz. KVKs, KGKs, TTCs, etc. are already equipped with video cassette recorder and monitors for use in routine extension programmes, itwould lessen the burden of initial investment. Extension personnel and farmers can be invited for training during off-season on a variety of practices of farming using live cases through video. It would avoid the cost of field visits and the need for synchronizing training programmes with the cropping season, thus, helping to carry out more programmes within the given resources. The author emphasises the need to develop interactive video for training farmers and extension personnel on crop production, poultry, dairy, sericulture and other farm related practices.

CONCLUSION

1. Specialists in conversation with farmers is the best mode of presentation for farm telecasts.
2. Farmers experienced difficulty in following the language was with technical words such as use of metric system of weights and measures. Also as many as 87% of the rural viewer complained frequent use of English words in dialogues
3. Farm programme producers in *Krishi Darshan Unit* should have enough background of rural life and interest in farming, so that they can perceive the information need of farm televiewers better.
4. Programmes in agriculture will have an immediate effect when they are telecast at the time when the farmers need them most during various crop stages.
5. The comprehension of the *Krishi Darshan Programme* can be increased presenting them with more illustrations, repetition of the salient points, and organizing farmers discussion.
6. Interestingly, farm women have a different choice on the timings of *Krishi Darshan Programme.* As high as 88% of them preferred the programme to be telecast from 8.00 PM onwards.
7. Several studies reported the gain in knowledge through farm telecast ranged from 43-73%. Hence, the duration of the *Krishi Darshan Programme* could be enhanced from the present half-an-hour to one hour.
8. With the lapse of time there is a reduction in knowledge gained Thus, it is suggested that the same programme may be repeated at the interval of 15 days.
9. Telecasts relating to demonstration of new skills and methods can best be presented through television as compared to radio.
10. TV exposures, supplemented with printed matter and group discussion will have better impact on information acquisition.
11. Farmers should be informed well in advance about the content of the forthcoming *Krishi Darshan* telecasts., If possible *Doordarshan* may prepare a simple folder containing the programme for each month and distribute it in advance to the TV villages.

12. Some meaningful entertainment is suggested in *Knshi Darshan Programme* for better reception of farm messages.
13. Majority of the farmers were found to be spending more time on discussing what has been seen on TV when compared to any other media. In light of the above observation, organizing tele-clubs is suggested in all the television villages.
14. A close collaboration among the television centres, State Agricultural Universities, ICAR Institutes and State department of agriculture is essential in farm programmes production, and
15. Recent advances indicate video can be effectively used in transmission of farm messages and video cassettes containing package of practices of various crops may be distributed to the enterprising farmers.

REFERENCES

1. *(1989). Impact of Agricultural Television Programme on Farmers in Delhi Villages* - Abstract m Evaluation *Report (mineo). Department of Adult Education, National Council of Educational Research and Training, New Delhi.*
2. *(1970). A Review of Delni Pilot Rural Television Project and its Lessons.* NASCOM REPORT. *Indian Space Research Organization and Department of Atomic Energy.*
3. (1986). Television in India, New Delhi: Audience Research Unit, Doordarshan.
4. *Chaturvedi, V.K. (1992). A study on Television Advertising; Farmer's Perspective.* Unpublished M.Sc. Thesis. *Division of Agricultural Extension, IARI, NewDelhi.*
5. *Chauhan, K.N.K. (137S;. inducing Change Through SITE; A Study* of *Some Socio-psychoiogicsi and Communication Correlates of Adoption Behaviour of the Rural Audience of SITE in North Bihar* Unpublished Ph.D. Thesis, *Dwision of Agricultural Extension, IARI, New Delhi-110012.*
6. *Chauhan, K.N.K. and Sinha. B.P. (1976). Impact of SITE on Adoption of farm technology,* Indian Journal of Extension Education, *Vol.Xlll.Nos.3 & 4:1421.*
7. *Chauhan, K.N.K. and Sinha, B.P. (1985). Inducing Change Through Farm Technology.* Ind. *J.* OfExtn. Edu. *VotXl!I.Nos.3&4;14-21.SITE: A Study of Some Socto-psyduilogical and Communication Correlates of Adoption Behaviour of the Rural Audience of SITE in North Bihar.* Abstract in Interaction. *Vol.3 No.si&i. 38.*
8. *Cherian, A. And CharxJra, A.* 113S3). *inpact ot Tefcv^nn on Acquisition*

and Retention of Knowledge Oy Rural People. Intf. *J.* Extn. EQu. *Vol.XXV,Nos.3S,4; 28-32.*

9. *Choudhary, A.K. (1986). Social Impact of SITE on Adults with reference to Agriculture and Animal Husbandry* Sampreshan. *National Council of Development Communication, Varanasi, .55p.*

10. *Cnowla, N.L (1934). A Perspective on Doordarshan in Mass Media in India.* Publication division, Ministry of Information and Broadcasting. Govt. Of *//ida, Patjala House, New Delhi, 1. PP. 13-17.*

11. *Dey, PK. (1989). Relative Effectiveness of Radk) and Television as Mass Communication Media In Dissemination of Agricultural Information.* Unpublished M.Sc. Thesis, *Division of Agnl. Ext., IARI, New Delhi.*

12. *Dubey, V.K. and SwamKar, U.K. (1985). Impact of Instructional Television Programmes on Knowledge and Attitude of Farm women in relation to Farm and Home Practices in Rajpur District, M.P.* Ssmpreshan. *National Council of Development Communication, Varanasi, PP-56.*

13. *Jha, R.C. (1978). A Study of Delhi TV: Retrospect and Prospect.* Unpublished *M.Sc.* thesis, *New Delhi-12.*

14. *Kamath, M.G (1973). Television and Social Change: Role in Television in Agriculture.* Agricultural Communication, *3: PP.19-22*

15. *Kamft, K. (1989). Growth of TV: The Crucial Role of Training.* Communicator. *Vd.XXIV No.2: PP.49-52.*

16. *Kashyap, P. (1994). TV 'Sky Invasion and Ground Realities'* Communicators *VOI.XXIX No.2: PP 5-9.*

17. *Kaur, R. (1970). Impact of Television on Farm Women* Unpublished *M.Sc. Thesis, Division of Agrtl. Ext., lARt, New Delhi.*

18. *Kishore Kumar, N. (1995). Video Technology and Rural Development.* Communicator. *Vol.XXX, No.1: PP -46-47*

19. *Kumar, G.A.K., Bhagat, R And Ojha, S.N. (1991). Television Viewing Behaviour - A Meta Analysis.* Ind J Extn. Edu. *Vol.XXVII, Nos.3S4: PP.5M2.*

20. Mahurkar, D.S. (1985). Television as Credible Source of Farm Information Sampreshan National Council of Development Communication. Varanasi PP.60

21. *Malhan, P.N. (1986). Communication Media - Yesterday, Today and Tomorrow.* Publication Division. Ministry of Information and Broadcasting. Government of India. *Patiala House, NewDelhi-1.*

22. *Mehra, 5., Sharma, M.C. and Nagar, S. (198S). Sociology of Television Viewing Behaviour.* Sampreshan. *National Council of Development Communication. Varanasi. PP.58.*

23. *Mehrotra,* N. *And Aujla, p. (1991). Impact of Television Viewing on*

Household Activities. Ind. J. Extn. Edu. *Vol.XXVII. Nos.3&4. PP 122-24.*

24. *Mishra, A.N . (1967). Impact of Television on Farmers.* Unpublished M Sc. Thesis *Division of Agril Ext. IARI. New Delhi.*

25. *Raghupathy, D. And Tatwar, S. (1990). Farm Telecast and small Fanners.* Communicator. *Vol.XXV, No.3: PP.3-5.*

26. *Ramesh Babu, A. (1976). Diffusion of Farm Information Through Television.* Unpublished Seminar report. *Division of Agricultural Extension, IARI, New Delhi.*

27. *Sadamate, V.V. (1976). Krishi Darshan Viewing Behaviour* of *Farmers and Informational Comparison of Television with other Sources of Farm Information.* Unpublished M.Sc. Thesis. *Division of Agril. Ext.. IARI, New Delhi.*

28. *Sadamate, V.V. and Sinha, B.P. (1976). Krishi Darshan Viewing Behaviour of Fanners and Informational Comparison of Television with Other Sources of Farm Information.* Abstract«Interaction. *Vof.3 Nos.152: PP.86.*

29. *Sadamate, V.V. and Sinha, BP. (1977). Farm Education Through Television.* tod. J. Extn. Edu. *Vol.13, Nos.3&4: PP.37-44.*

30. *Sohl, P.NP. (1977). A Study on Audtence Profile of SITE and Its Tecnno-Rural Impact in Goodak Command Area of North Bihar* Unpublished Ph.D. Thesis. *Division of Agrl. Ext. IARI, New Delhi.*

31. *Sangha, G.S., Singh, N.P. and Gupta, MP. (1979). Awareness of the Rural TV viewers about the Developmental Programmes of Amritsar Doordarshan.* Ind. *J.* Extn. Edu. *Volis, Nos.384: PP.74-76.*

32. *Sangha, G.S. and Gupta, M.P. (1986). Credibility of Television as a Source of Information.* Ind. J. Extn. Edu. *Vol.XXI, Nos.182: PP.74-76.*

33. *Sadangi, B.N. and Rajguni, G. (1985). Impact of SITE Programme on Adoption behaviour of Farmers.* Sampreshan. *National Council Development Communication, Varanasi. PP.57.*

34. *Sekhon. I. (1968). A Pilot Study of Assess the Effectiveness of Television as a Medium of Communication for Imparting Technical Know-how to the Fanners.* Unpublished M.Sc. Thesis. *Lady Irwin College, New Delhi.*

35. *Sekhon, L (1970). The Effectiveness of Television as a Medium of Communication for Imparting Scientific Know-how to the Fanners.* Ind. J. Extn. Edu. *Vol.VI, Nos.1&2: PP.90-96.*

36. *Shwrna, M.C., Mehra, 5. And Kaur, P. (1985). Mass Media and Change with Special Reference to Television.* Sampreshan. *National Council of Development Communication, Varanasi. PP.54.*

37. *Snaima, S.K. and Dey, PJC (1970). Relative Effectiveness of Radio and' Television as Mass Communication Media' in Dissemination of Agricultural Information.* InA J, Extn. Edu. *Vol.VI Nos.1&2: PP 62-67.*

38. Singh, B. (1992). Impact of Television on the Indian Society, / nteracfcv).

VolJX, No.1: PP.134-60.

39. *Singh, J. (1971). A Study of the Factor Influencing the Viewing Behaviour of the Farmers Towards Agriculural Programmes on Television in Delhi Village.* Unpublished Ph.D. Thesis, *Division of Agril. Ext., IARI, New Detii.*

40. *Singh, IB. (1971). Satellite Television for Indian VSages - Prospect and Retrospect.* Ind. J. Extn. Edu. *Vol.VII, Nos.3&4: PP. 138-43.*

41. *Singh, UP. and Prasad, C. (1976). Content Analysis of Farm Telecast and Viewers Information Need.* Ind. J. Extn. Edu. *Vol.XI, Nos.1&2.*

42. *Singh, N.P. (1977). A Study of the Factors Influencing Farm Televising Behaviour Among Delhi Farmers.* Ind. J. Extn. Edu., *VoUCUl, Nos.1&2; PP.15.*

43. *Singh, R. (1971). A Comparative Study or Adoption of High Yielding Varieties Programme by the Farmers of Farmers and Non-Television Villages.* Unpublished M.Sc. Thesis, *IARI, NewDelhi.*

44. *Singh, R. And Sharma, S.K. (1973). A Comparative Study of Adoption of High Yielding Varieties Programme by the Farmers of Television and Non-Television Village.* tnd. J. Gxtn.Edu., *VoJ.IX, Nos.3S4: PP.92-101.*

45. *Singh, Ft (1985).* Transference of dairy innovations through Television. *Sampreshan, National Council of Development Communication, Varanasi, PP. 59.*

46. *Singh, Y.P. (1990). Farm Message to Millions: The Indian Studies.* Interaction. *Vol. VCNI, No.2PP.1-63.*

47. *Sinha, BP. (1970). A Study of Some Motivational Factors in Diffusion of Farm Information Through Television. Ph.D.* Thesis, *IARI, New Delhi-12.*

48. *Sinha, B.P. and Mehta, P. (1971). Fanners Motivational Dispositions and liking for Krishi Darshan Programme of Delhi Television.* Ind J. Ext Edu, *Vol. VIII, NOS.3S4: PP.6&-73.*

49. *Sinha, Bf. (1974). Diffusion of Information from a Farm Telecast,* tnd. J. Ext. Edu. *VoJJX, Nos.18.2: PP.66-70.*

50. *Sinha, B.P. (1974), Television in Diffusion of Farm Information: A Motivational Study,* Delhi. New HgMght.

51. *Sinha, B.P. (1985), TV in India: Shall Ruralities have their Due.* Sampreshan, *National Council of Development Communication, Varanasi: PP.53.*

52. *Subbaryan, G. (1991). Television and Social Behaviour.* Communicator, *Vol.XXVI,No.4: PP.40-41.*

53. *Talawar, S. (1994). Interactive Video in Training,* Conrnunicatof. *Vol.XXIV, No.1:PP.5-7.*

54. *Wilber Schranwn (1953). What may have realistically expect of Educational Television?* In Cooperative *Ertenswn* Work. *Keteey, L.D. and Heame, C.C. Ithaca, New York: Comstock and assoc.*

Chapter **23**

Communication Research and Agricultural Development Present Status and Future Strategy

Ranjit Singh

Since independence special emphasis has been laid on agricultural development as country was facing acute shortage of food-grains and majority of the population has farming as main occupation. Since the launching of Grow More Food Campaign in 1950, a large number of innovative communication techniques were employed to bring desirable change in the knowledge, attitude and skill of the farmers of the country. Although, most of these techniques have many planning and implementing drawbacks yet they showed good results. The country increased its food production from 80 million tones to nearly 180 millions tones. The early days of Green Revolution raised the level of aspiration of farming families for better life specially of younger generation. Anyhow over the period of time, the margin of profits has been decreased and the production level is almost static for the last decade. This has resulted in some sort of frustration among the farmers. On the other hand, the growth of the population in India is taking place at a very high rate. It is estimated that the population of India will exceed one billion at the turn of the country. The food requirements at that time would be about 224 million tones. It the seed requirements and the loss during storage and handling are added, then India will need 252 million tones of food grains. At present the total food-grains production is about 180 million tones. Thus, an additional 70 million tones of food-grains would be required to meet the minimum food requirements of our population. The sharp increase in population has also reduced the per capita availability of land, which is now one of the lowest in the world. The availability of food per capita is also low in the country. Intensive efforts are required in this decade to increase food production in order to meet the minimum food requirements of our population. New opportunities have also been made available due to globalization to trade. In fact, globalization has brought forth opportunities as well as threats. We have to improve the quality and bring down the cost of production in order to compete in the world market. Farmers can earn more by producing quality products for the world market. Farmers can earn more by producing quality products for the world market. We have to do everything possible not to defend and enlarge the gains already made out to provide new direction to the agriculture of future.

In order to produce for the world market, a dynamic agricultural production system is required. This will require suitable technology, effective communication and delivery system and supportive Government policies. High tech. Agriculture involves more precision, higher risks and active involvement of the farmers. In fact, it is the best time for the farmers but at the same time it is the worst of time for farmers. It is the best time for those farmers who can produce for international market and it is the worst time for those farmers who will not be able to complete in the tough competition. We have to think globally considering the whole intellectual challenge but have to act locally to tackle the specific elements which can be changed. The success of high-tech. agriculture depends upon :

Efficient technology generation system :

1. Effective technology communication system.
2. Adequate services and supplies system.
3. Reliable quality assurance system.
4. Suitable production management system.

Above all, there is a need for Responsive Farming Community.

It is a fact that as the modernization of farming techniques takes place, the intensity of risk increases. It has also been seen that the effectiveness of traditional communication system decreases with the progress of agriculture. The innovative farming community required specialized and energetic communication system with quick response to farmers' queries. There is need to change the perceptions of all concerned as our world is made up of our perceptions and nothing else matters. Since the risk intensity increase the risk taking capacity of the farmers. Before discussing the required communication strategies and research system for developing the same, it is essential to analyze the existing techniques so as to suggest desired changes.

An analysis of the existing technologies can suggest following short comings :

1. Most of the available transfer technology systems for family and community improvements are fragmented and limited in scale.
2. There is no coordination between different programmes. It is common to find in the same rural area series of small separate programmes for health, nutrition, home economics, family planning, cooperative, sports, recreation, etc. All aim at the same audience yet sponsored and operated by different public and private agencies with little if any co-ordination or cooperation.
3. These programmes are often of poor quality and spread so thin that they have little impact.
4. There are very few programmes for farm women. Although, they play much more important role at farm and home.
5. General education and operational skill training programmes for out of school youth are almost non-existent in rural areas. The few non-formal programmes for young people often benefit only those who are still attending school.
6. There is no agency or department responsible for maintaining the over-view of all programmes for projecting future needs or acting as coordinator.
7. These programmes are operating hap-hazardly with neither priorities nor plans.
8. It has been observed that most of the agricultural communication programme concentrate their efforts on larger producers and major commercial crop while neglecting small farmers and the local crops of key importance to subsistence families.
9. Very little efforts has been made to diagnose the different needs of client families and standardized recommendations has been handed out that many of these farmers find impracticable and useless.
10. There are not inadequate facilities for in – service and refresher training to the field staff.
11. Most of the time field staff has been burdened with responsibilities which are non-productive.
12. Field staff has not been provided with adequate transport and other facilities.

13. Field efforts are not adequately reinforced with mass media and other communication supports.

Earlier communication process was a one step flow of information from extension agency to the farmers. Later on, it became two step flow i.e. from research to extension agency and from extension agency to the farmers. Presently, it has to be direct and effective interaction between research, extension and farmers. In this system, farmers have to show their needs and concerns and have to decide for collective action. The required knowledge input and other facilities have to be provided by the research and communication system. Collective action is required because people are more willing to accept risks if they have participated in deciding what risks are acceptable and there are alternatives and they are clearly understood. The experts are competent and trusted and are seen as members of community. Messages must include a combination of statistics and experiences. Any development strategy which aims at improving the lot of the farming community must aim at creating new productive assets for them. The emphasis should be to help them in order to help themselves. A successful community system is that which is able to create awareness among clients and unite them to solve their problems collectively. Suitable support in terms of providing suitable technology, marketing facilities, training and up-gradation of skill has to be provided to make the programme a success. Foremost necessity is to evolve suitable technology followed by effective communication strategy and required infrastructure. Anyhow, it is of no use to improve the knowledge base for a particular activity, without examining the specific work situation and the degree of freedom available to act on that information. Programmes should be those which help people to stand on their own feet, that genuinely encourage responsibility, initiative, decision making and self-reliance.

Research in development communication :

We have to admit that research in communication mainly tended to be practice-oriented. As a result, theory-building has been neglected. Practice-oriented research aimed at solving field problems is not unimportant or unnecessary, the problem is a lack of balance between practice-oriented research has become almost a theoretical due to lack of inter-disciplinary orientation. Most of the research studies during the last three decades has been concentrated on the following process :

1. Information sources at different stages of adoption process.
2. Key communicators.
3. Impact and content analysis of mass media.
4. Information needs.
5. Modes of communication.

It is also a fact that we have not been able to add methodological sophistication to research. There has been a tendency to rely heavily on the traditional quantitative methodology without necessarily ensuring the methodological rigour that accompanies its use. While there is evidence of the use of qualitative methodologies as well, there is still scope for using more innovative research methodologies.

Important issues :

Some of the important issues in communication research can be identified. A discussion of each issue will help in having a clear picture of the present communication research along with the problems relating to each issue.

1. **Status of research :** While discussing present status of research in communication it can be stated that there are very few research schemes with whole time researchers. At present, coordinate and collaborated research projects are non-existing and not much significance has been attached to inter - disciplinary approach. In fact research in communication and extension has rarely get any importance thus, no body provides resources and facilities. The obtained results have rare applicability.
2. **Selection of research problems** : As already stated most of the research studies are replicated and are of stereotyped nature. In fact, research has not gone beyond the above listed five topics. Therefore, selected topics are insensitive to problems being faced by farming community and development agencies. Since majority of the research studies are conducted by graduate students, they have little access to latest literature and remain confined to already completed theses. As there is little interaction with development departments, our studies are less realistic and are not need based.
3. **Methodology used :** Over the period of time it has not become possible to add sophistication to research techniques and majority of the studies are confined to the use of sample survey technique. Research workers has not been adequately trained

in instrument development and data collection. They use faulty instruments and collect non-reliable and of poor quality data. The responses given by the respondents lack authenticity, quality, validity and genuineness. Respondents also do not fully cooperate and provide incomplete and non-reliable information. Infact, there is methodological inertia and stagnation.

4. Research results – Results are rarely shared with concerned organizations. Even if presented people do not take them seriously. Since most of the studies are conducted by students with limited resources, thus, results lack authenticity, reliability and quality.

As in other disciplines, there is a need for collaborative and coordinated research studies with the help of inter-disciplinary research team. All India Coordinated Research Studies should not only improve the quality of research but also contribute to theory-building. Cross – cultural comparative studies would also help in improving the quality of research. There is also need for proper dissemination of research findings to improve the performance. Most of degrees, findings of these studies are neither documented systematically nor disseminated effectively among researchers, practitioners, policy makes and planners.

Projections for future research :

It is important to decide the Direction of the research in future. Some of the suggestions are given below :

1. Developing media kits for different situations and different categories of farmers.
2. Information strategy for different situations.
3. Encouraging research in dialogical and risk communication.
4. Research in developing multimedia campaigns.
5. Promoting use of promotional and folk media.

There is also a need for an institute or centre that would energize, catalyze and coordinate research. Such an institute could provide academic leadership to various institutes and agencies. There is need to add sophistication to the planning of information management. Information strategy planning for different situation with the help of computers is required. Information strategy planning focuses on the strategies business goals, identifies the key information needed to successfully direct the farmers and formulates future plans of information management.

Attention has also not been paid for evaluation of the outcome of various communication approaches resulting in quick changes of approaches without caring for improving the existing approach. Not much research work has been done concerning testing of various theories and models of motivation. It is usually considered that motivation depends upon individual innovation and situational characteristics. The performance and satisfaction depends on the individuals' compassion of his or her contribution and rewards with those of other in similar situations. Whereas the expectancy model is based on motivation, performance, satisfaction and on what the individual experts from the proposed performance. There is need to develop and test various models of communication which can be applied under different situations.

The use of field experimental design by communication researchers to study development problems moves research towards development programmes. It puts the communicator in the role of development designer as well as that of research evaluator.

There is also a need to focus on interpersonal networks. Network analysis will tell us the exact nature about from where opinion leaders obtained the message and specifically to whom each such opinion leader disseminated the message. Locating structural constraints to development is another important subject of inquiry. More specific questions must be asked about development or impediment to development. Instead of assuming general sequences that are good for modernization, we must investigate the possibility of different variations of those patterns. We must assess hard fact on the conditions that bring about desirable change in a given situation and develop some general principles.

Shifting emphasis of research :

Based on the above discussion the emphasis of research should shift as per suggestion given below :

1. Shifting our attention from the end users to the decision making process.
2. Communication role in the structural change processes.
3. Shifting attention from individual effects to institutional effects.
4. More emphasis on improving quality of products and labour.
5. Specialized, energetic and computerized technology transfer system rather than general communication.

Some of the new technologies can be identified as under :

1. Microwave transmission
2. Local transmission
3. Local cable transmission
4. Computer transmission of documents/data
5. Two way compressed video transmission using CODELS.

New communication technology :

Revolutionary changes have occurred in the field of communication. New technologies offer us an unprecedented opportunity to accelerate the transfer of knowledge and to provide wide access to information. Wisely used, they can be a powerful tool for realization of our goals. However, concerted efforts and considerable resources are required to make use of this technological revolution. We have to change training contents for preparing our communicators capable of handling new information technologies.

Presently, professional communicators produced the messages for the audience. The audience remained powerless to respond directly to or interact with the message senders. By contrast, the most revolutionary feature of the new means of communication is that many of them are interactional permitting each individual user to make or send images as well as merely to receive them from the outside. Two way cable video cassette, cheap copies and tape recorders, all place the means of communication in to the hands of the individual. Looking ahead, one can imagine a stage at which even ordinary television becomes interactive.

A greatly enlarged supply of learning aids specially print materials, radio and television programmes can spread the message more effectively, quickly and accurately. There is need to pay more attention to create good educational software for both print and electronic media. Market places and other popular meeting places can be used for creating successful teaching - learning situations. Traditional forms of folk entertainment can be employed as communication channels. The educational programems should not only concern with teaching and securing the adoption of a particular improved practice but also concerned with changing the outlook of the farmers and encouraging their initiatives in improving their farms and homes. In fact, it has to be particularly communication. Emphasis has to be laid on specialized training and whole family. It is also essential to prepare media kits for different situations and different categories of the farming families. Without an appropriate communication strategy, deliberately planned, organized and directed in a manner and form popularly relevant to the existing environment and the desired change, very little in development terms is likely to be achieved.

Blending new communication activities with traditional forms, rituals, festivals and educational institutions, in terms of the norms and practices of those institutions, are important for encouraging local initiatives and utilizing local talents and resources. The mass media development communication programmes need to do more in promoting discussions and debates various groups to promote dialogical communication, rather than reporting group activities and presenting generalized packaged information, which encourages monological communication.

To ensure that the media make an effective contribution to promote adoption of modern techniques of farming, it is necessary to adopt what is called a multimedia approach. It means utilizing all the communication media plus the interpersonal communication channels available for designing an integrated communication programme. This approach saves the efforts, avoids wastage and provides stronger, more coherent and more convincing messages. A multi-media approach makes the most rational and economical use of all available communication resources so that each reinforces the other. Different media have strength and weaknesses. Although broadcast media may have the greatest frequency and reach among a particular audience, print materials can provide detailed information and can be kept in the home to be used when needed. Face to face interaction can be provide personal reinforcement not available through other media. Community networks can provide extensive and deep support. It is, therefore, important to carefully select elements from each of the media groups and then integrate them so that they multiply the important of each other. The reach of broadcast messages combined with the detail and relative performance of print followed by training programmes provide opportunities to reinforce complicated messages. Mass media, folk media, promotional media and associated or supporting media are the possible components of a multimedia programme. Correspondence course, audio cassettes, crop calendars, vernacular newspapers, video films, posters etc. if effectively prepared and properly used can form important components of multimedia campaigns. Mass media can play a significant role in creating awareness, convincing and prompting action. Folk media are very useful for developing faith in science and technology.

Promotional media give additional support to the campaign. They include such attractive and pleasure producing devices as badges, tee-shirts, mementos, etc. indicating that person has modernized his farming. Associated media are means of communication that are used in association with the other media. They include such materials as pamphlets, display charts, slides, flip charts, tapes and cassettes for use either by an individual alone or in the training camps and field days. Associated and supporting media materials might be distributed to the participants of training camps and field days. The computers can play a very significant role in developing print and supporting media of high quality. While developing a multimedia programme for promoting adoption, it is important to know about the target groups. The message that is to be communicated to each group should be designed keeping in view their present level of adoption and available resources. After studying target groups and finalizing messages, it is important to identify the media for the campaign along with role of each medium and the possible media approaches. A working plan should be prepared by the concerned department in consultation with the representatives of the various media. As already stated, communication strategy needs drastic changes. There is a need for an effective interaction between research, extension and farmers with efficient organizational support. Communicator has to be an action researcher while advocating new technologies to the farmers. Action research integrates theory and practice in a manner that has a profound effect on the way we view the world. From this perspective, the communicator is not a mere transmitter of knowledge, but an action researcher. This professional research locates issues of relevance to his or her client farmers with the farmers themselves participating in the research and in so doing, learning the ways of inquiry and interacting with the scientist along with the farmers to seek new technologies and information. As previous studies have confirmed, special attention need to be on educated rural youth for high-tech agriculture. Education and literacy are pre-requisite for adopting sophisticated technologies. Moreover, youth have high risk taking capacity. It will also help in minimizing the inflow of rural migrants to cities. Many rural young people have no idea of what to expect before they make move to the cities. If they can have better opportunities in villages, many of them may like to stay back in the village. These young farmers will not only act as key communicators but will also be active partners in the participatory research. In fact, success of farming for world market will depend upon the active collaboration of research, extension, farmers and Government from the decision making stage to the final execution of the project. Any project finalized with and by the people stress the principles of access, participation and self-management, provides as efficient mechanism for information seeking and feedback. We should study how communication can be effectively used to make grass roots contribute to the decision making process. Communication has to play an important part to make sure that the decision making process is not faulty, or atleast contains some internal mechanism that can avoid or minimize the chance of disaster in the high risk farming.

We should shift our attention from individual effects to institutional effects. There are very few efforts that look into the effects on institutions or on the social structure as a whole. Collective efforts and development of institutions are required for the success of high-tech farming. The requirement of labour will be of different nature requiring new labour management techniques. There is also a need to know what institutional changes take place in order to facilitate the adoption of high-tech agriculture.

There is need to use modern communication methods. The use of modern communication technologies to educate the farmers required the transformation of entire social structure. Adoption of modern technologies of communication and production involve manpower quality, work ethics, wages and incentives, management style, transportation, marketing, consumption, capital accumulation, credit and banking etc. Unless effective institutions develop to take care of these and other important functions, it will not be possible to absorb the modern technologies of either communication or production. If we have to produce for the world market we have to modernize ourselves. Only then we can introduce modernization among our farming community. The concept of social marketing has not been utilized in communication process. The design, implementation and control of programmes seek to increase concepts of market segmentation, consumer research, idea configuration, communication, incentives and exchange theory to maximize target group response. Marketing is a process of deciding what the best mix of product, price, place and promotion is. The definition of social marketing has been stretched in recent years to include not only the marketing of products, but also attempts to persuade people to adopt a behaviour or accept a socially beneficial idea.

Failure of programmes may be due to many reasons but one which has been largely overlooked is the disparity between the way Government bureaucracies operate and the requirements of development programmes which can be mobilize the rural poor for sustained, effective self development and encourage innovative farmers for adopting high-tech agriculture. Too often overlooked is the reality that the response of the people to Government programems is shaped by the way of services are administered, how accommodating or inflexible the services are, how satisfying or how humiliating their treatment is, how readily the farmers get access to services or how much more readily the influence can utilize them, whether Government staff adopt problem-solving stance or in conventionally bureaucratic one, how attorned staff are the actual conditions and needs of the people, whether these staff deal with them as responsible adults or as basically ignorant and irresponsible adults. If we want to achieve equitable, self-sustaining improvements in the conditions and capacities of the majority, then we must find way to make the long term commitments necessary to achieve the institutional changes in bureaucracies which are basic to that outcome.

In future, our emphasis will not only be on quantity but quality also. Quality is the most potent weapon to face the fierce competition and it needs to be nurtured through the right attitude and a quality culture. In the twenty-first century, doing things right or quality would not only be a plus point, it would be an essential requirement for survival. Traditionally, quality means conforming to specification and standards. But in this changed scenario due to WTO, quality also means productivity, competitive costs, timely delivery and total consumer's satisfaction. If farmers are inspired by a desire for change, they will have the necessary energy, determination and commitment and will take stock of the means, resources and skills required. Unless a man feels the desire to have more material wealth, he cannot be expected to have much interest in new techniques. A powerful and innovative communication programme can play a significant role in motivating the masses for adoption of new technology to produce more and of high quality. Anyhow, it has been observed that communication is still not seen as a powerful force for change or as a means of passing the know how more economically and effectively, because most of the projects are usually planned by technical specialists who are not trained in development communication. It has to be admitted that communication plays a significant role in the agricultural development. But that can only be possible if an appropriate communication strategy is planned and implemented. There is a very strong correlation between the economic health of a nation and the learning health of its citizens. In an age in which information doubles every five years and then feeds upon itself to produce new knowledge a strong communication system is essential to keep the society well informed. An appropriate communication strategy based on multimedia approach deliberately planned, organized and directed in a manner and form popularly relevant to the changing environment and the desired change will be essential for developing agriculture in the next century.

Chapter **24**

Farmers Perceived Usefulness and Quality of the Farm Page of an Assamese Daily

U. Barman, D. K. Gogoi

Newspaper is one of the important mass media in the world. A wide variety of topics are covered by newspapers and serve a large number of readers with varied interests. To day, news paper is read by majority of the literate villagers since it has started reaching in remote villages also. A properly edited, printed and regularly produced newspaper may serve to establish a continuous link with the people by giving education as well as relevant and timely information on many aspects of farm and home (Sawant et. al. 1986). If such a newspaper covers agricultural topics, it's text should be readable and easily understood by farmer readers. At present, many of the Assamese dailies have started publishing farm page. In fact, little effort have yet been made to make them effective tools of communication. It becomes necessary to collect experimental evidence to know how does the farmer view on the usefulness and quality of the content covered by the farm pages. The analysis will help to make the farm pages more effective. Keeping in view, the present field level study was designed to study the perceived usefulness & quality of contents of farm page of highest circulated Assamese daily viz. 'Dainik Janambhumi'.

Research Methodology

The present study was conducted in Jorhat District of Assam because the literacy rate of the district was highest in the State, and the circulation of the newspaper is also highest in the district. Five numbers of Rural Agents of the newspapers were selected at random from all the three Agricultural Sub-divisions of the district. A sample of 150 reader farmers were selected randomly from each of the selected Agents with proportionate allocation. The respondents were personally interviewed with the help of well-structured pre-tested interview schedule developed for the study.

Findings and Discussion

The respondents were grouped into different categories on the basis of personal characteristics. It is evident form the Table-1 that majority of the respondents (47.33%) belonged to middle age group ranging form 36 to 50 years followed by old age category (32.67%). One third of the respondents (33.33%) had graduate or above education followed by high school level education. It was found that the majority of the respondents (45.30%) were big farmers followed by 44% small farmers. It was found that 65.30% and 57.33% respondents had medium level of economic motivation and innovation proneness.

Table-1 : Distribution of respondents according to their socio-personal characteristics (N=150)

Characteristics	Category	Frequency	%
Age	Young (below 36 years)	30	20.00
	Middle (36-50 years)	71	47.33
	Old (above 50 years)	49	32.67
Education	Primary school level	14	9.34
	Middle school level	20	13.33
	High school level	30	24.00
	Higher secondary level	30	20.00
	Graduate level and above	50	33.33
Size of operational land holding	Marginal (upto 1 hact.)	16	10.70
	Small (1-2 hact.)	66	44.00
	Big (above 2 hact.)	68	45.30
Economic motivation	Low	15	10.00
	Medium	98	65.30
	High	37	24.70
Innovation proneness	Low	30	20.00
	Medium	86	57.33
	High	34	22.67

Perceived usefulness of the content

The table-2 shows that a large proportion of the respondents (83.33%) perceived that the contents are published timely. It may be concluded that the farm page communicate the message to the farmers in time. Regarding applicability, 54.67% respondents found that the contents were partially applicable for them. More than half of the respondents (58%) mentioned that the contents partially solved their farm problems. The data indicated that 62.67% respondents perceived that the contents partially fulfilled their needs related to farming. Thus, it can be concluded that the agricultural information published in the daily are partly need a based, applicable and help in solving the day-to-day problems of its reader farmers. The table-2 indicates that one-third of the respondents (74%) state that sometimes new farm information are published in the farm page. This may be due to lack of strong linkage of the daily with Agricultural Departments/ University.

Table-2 : Perception of respondents about selected attributes of usefulness of content published in the farm page (N=150).

Attitude	Category	Frequency	%
Timelines	Timely	125	83.33
	Not timely	25	16.67
Applicability	Highly applicable	65	43.33
	Partially applicable	82	54.67
	Not applicable	11	7.33
Needfulness	Rarelly fulfilled	38	25.33
	Partially fulfilled	94	62.67
	Not fulfilled	18	12.00
Regularly of publishing new farm information	Regularly published	39	26.00
	Sometimes published	111	74.00
	Not published	00	00.00

The interval level data on perceived usefulness of the content (Table-3) showed that 63.33% respondents perceived the usefulness of the content as medium followed by 32.67% with high level of perception.

Table-3 : Distribution of respondents on the basis of overall perceived usefulness score (N=150).

Category	Frequency	%	Mean	Standard Deviation
Low	6	4	-	-
Medium	95	63.33	6.71	1.95
High	49	32.67	-	-

Perceived quality of contents of the Farm Page

The table-4 reveals that a major proportion of the respondents (84.69%) assessed that the content published in the farm page were easily understood by them. According to 87.33% respondents, the content had no difficult Assamese words. On the attribute of "use of unfamiliar English words" 40.67% respondents stated that the writers used unfamiliar English words in the content. The reason may be difficulty in writing on scientific farm information without using English words. The table-4 indicates that 49-33% respondents mentioned the content some what explain the importance of practice recommended for them. Only 3.34% opined that it was not explained properly. Fifty percent respondents revealed that content somewhat explains the technology advocated for them. It concluded that proper emphasis was not given on explaining the importance of the recommended practices and also on how to adopt them. The contents covered few relevant points as stated by 56.67% respondents. It indicated that the contents published in the farm page do not cover all the pertinent fact of an idea communicated to its farmer readers. The contents of the farm page were perceived to be easily readable by 86.67% respondents-Majority of the respondents (92.67%) were satisfied with the present style of arrangement of the articles. The table-4 also indicates that 68% respondents are satisfied with the present pattern of using photograph and diagrams. Thus, it may be concluded that the farm page published almost in an attractive way but there is still scope for inclusion of more photograph and diagrams etc.

Table-4 : Perception of respondents about selected attributes of quality of content published in the farm page (N=150).

Attribute	Category	Frequency	%
Understandability of the content	Understand easily	127	84.67
	Understand with some difficulty	23	15.33
	Not understand	00	00.00
Use of difficult Assamese word	Not used	131	87.33
	Used but few	19	12.67
	Frequently used	00	00.00
Explainariability of importance of practice recommended	Properly explained	74	49.33
	Some what explained	71	47.33
	Not explained	07	4.67
Expainability of tech. advocated	Properly explained	75	50.00
	Somewhat explained	65	45.33
	Not explained	07	4.67
	Not satisfied	11	32.00

Attribute	Category	Frequency	%
	Somewhat explained	65	45.33
	Not explained	07	4.67
Relevancy of the content	Covers sufficient relevant point	64	42.67
	Few relevant points	65	45.33
	No relevant points	01	0.66
Readibility of the content	Easy reading	130	86.67
	Read with some difficulty	20	13.33
	Could not read properly	00	00.00
Arrangement of articles	Satisfied	139	92.67
	Not satisfied	11	07.33
Publication of photograph/Diagram	Satisfied	102	60.00
	Not satisfied	11	32.00

The findings of interval level data analysis for perceived quality of the farm page is presented in table-5. The findings revealed that 90% respondents perceived the quality of the farm page as high. Though the overall perceived quality was high but some attributes related with subject matter of the contents like "explainabllity of importance of practice recommended" ... "explainability of technology advocated" and "relevancy" have to be improved. Because half of the respondents had low to medium level of perception on these three attributes.

Table-5: Distribution of respondents on the basis of overall perceived quality score (N=150)

Category	Frequency	%	Mean	Standard Deviation
Low	00	00.00	-	-
Medium	15	10.00	16.6	2.31
High	13.5	96.00	-	-

Conclusions

The study revealed that the farm page of 'Dainik Janambhumi' was moderately useful for majority of the reader farmers. Though the perceived quality of the farm page was high, there is still scope for improvement. So, measure should be taken to publish need based topic of high applicability. The contents should help in finding solution to most of its readers. Latest farm information covering all the relevant points should be published regularly. For this purpose, the publisher may establish a linkage with agriculture and other allied departments and Assam Agricultural University. This will help to improve the usefulness and quality of the farm page.

REFERENCE

1. Sawant, G.K.; Shendre, N. K. & Pawar, M.D. (1986) Textual Characteristics of the farm information items appeared in the leading Marathi newspaper in Maharashtra. Interaction 4 (182): 51-62.

Chapter **25**

Textual Characteristics of the Farm Information Items appeared in the leading Marathi Newspaper in Maharashtra

G. K. Sawant, N. K. Shendre and H.D. Pawar

Farm information items appeared in leading **Marathi** newspaper in Maharashtra were analyzed. Most of the farm information items were on right side of the middle pages of newspaper. Flush left and cross line types of headlines were common in most of the farm information items. Most of the farm information items had heading and body but rarely lead and illustrations. The size of the farm information items was upto 200 words. Most of the farm information items had more than six per cent difficulty words, 1 to 6 per cent technical words, two per cent personal words. The average length of sentences ranged from 8 to 17 words per sentence. Majority of the farm information items had 10 to 20 sentences, fifty per cent simple sentences and compound sentences upto 50 per cent. Most of the headings and leads of farm information items had one sentence and 10 sentences in body.

Among the mass media used in a country like India, newspaper occupies a place of prominence because of its low cost per copy, high circulation, availability and **variety** of contents. Newspapers not only inform, educate, and entertain the readers but **also** motivate them for action. It also helps to sustain literacy in **case of** neo-literates. Today, newspaper is read by majority of the literate villager; (Sawant and Waghmard. 1985) Newspaper have started reaching in remote villages (Joshi and Kuikarni 1978-79). Though agriculture is regarded as national industry, agricultural news were not properly covered in newspapers (Mathur, 1984).

A properly edited, printed and regularly produced newspaper or periodical may **serve** to establish a continuous link with the people by giving education as well as relevant and timely information on many aspects of the day-to-day life. If such a **newspaper** covers topics of rural matter and farm information items, its text should be readable and easily understood by the farmer readers. With this objective in mind, the present investigation was carried out to study the textual characteristics of the farm information items appeared in the leading **Marathi** newspaper from Maharashtra.

Methodology

Terms and concepts used in this investigation.

I. **Farm information item :** Any piece of text, illustration, or any other form of presentation appeared in the newspaper relating to soils, crops, animals, poultry, beekeeping, horticulture or forestry. Where these have economic value in the day-to-day farming, agricultural policies or agricultural activities of government agencies or organizations.

II. **Leading newspaper :** The leading newspaper is one which is distributed all over the state, possesses consistency in publication and highest in circulation.

III. **Newspaper :** Newspaper is a daily publication which covers news events, feature articles, sprots roundup, editorial, market prices etc. advertisements for the use of readers.

IV. **Textual characteristics :** These are the physical dimensions of text, which makes text meaningful and understable, following textual characteristics were considered.

1. Location: It refers to the position of agricultural information item on number and side of page.
2. Type of Heading : The type of heading is decided on the basis of arrangement and position of heading in the agricultural information item with respect to a column. There are several types of heading : Flush left, Crossline, Flush head, Pyramid, Inverted Pyramid, Kicker, Reverse Kicker, Hanging.
3. Lead : It follows immediately after the heading. A line of text that expands the thought given in the heading, clarifies it or adds information (Holtje, 1978).

4. Body : The body of the agricultural information item is the main text, which describes the event in detail and differentiate from the heading and lead by smaller type (Holtje, 1978).
5. Size of text : It is tonal number of words in the body including heading.
6. Difficult words : Difficult words are those which are hard to read, write and understand.
7. Simple words : Simple words are those which are easy to read, write and understand.
8. Personal words : These are the words which are referred to person to develop direct intimacy with reader.
9. Technical words : These are the words which describes some technical know-how, procedure, phenomena and object.
10. Sentence : A sentence is a group of words arranged gramatically that makes meaning.
11. Simple sentence : A simple sentence is a meaningful arrangement of words, which conveys one idea.
12. Compound sentence : A compound sentence is one that comprises of two simple sentences and conveys more than one ideas through it.
13. Illustration : The illustration in the agricultural information item is a graphic presentation like photograph, drawing, which are used to portray the agricultural information and attract attention and promote interest in the information item being communicated.

Selection of Newspaper : All newspapers published in **Marathi** language from Maharashtra were considered. A daily newspaper in Marathi language **-Sakal-** was identified as a leading newspaper. Ten per cent (Thirty six), issues of the newspaper of the year 1986, three issues per month were randomly selected for the purpose. Farm information items appeared in the selected issues of the newspaper were analyzed for their textual charcteristics and were expressed in terms of frequencies and percentages.

Results and Discussions

It is a common observation that alongwith non-agricultural information items, agricultural, rural development matters and advertisements appear in newspapers. It was observed that slightly more than one third (38.6896) newspaper space was allotted to the non-agricultural information items andagricultural information items occupied only 2.32 per cent newspaper space. Advertisements occupied 59 per cent space in newspaper under study.

1.Position of Agricultural Information Items

The position of the agricultural information items on different pages of the newspaper was studied. It was found that nearly thirty per cent *(29.16&)* farm information items appeared on the fifth and slightly more than twenty per cent (23.90%) farm information items appeared on eight pages. This may be due to the fact that in order to catch up urban readership editors may be placing the farm information items on middle pages. Non-agriculturalinformationitems might have given space on first page of the newspaper. It was observed in the present study that only 5 per cent (5.37%) agricultural news items were published on the first page. Most of the agricultural information items appeared on middle pages (3rd, 5th, 8th & 11th) of newspaper.

2. Location of Agricultural Information Items

It was observed that the maximum number (62 per cent of the agricultural Information items appeared on the right side of the newspaper and remaining (28 per cent) on the left side.

3. Headings in Agricultural Information Items

Types of headings in the agricultural information items were studied. Among the different types of headings, flush left type is easier to write and it often tells more about the news and it was used in about twenty five per cent (24.39%) of the agricultural information items. The crossline type of heading was generally used for short stories (Crump. 1974). But most of the agricultural information item of the newspaper under study were short. Therefore, crossline type of heading was seen in about 22 per cent of farm information items. The other types of headings like flush head, pyramid, kicker etc. were few in number. The type of headings like, middle column, one column heading two column news, four column heading two column news etc. were also observed.

4. Heading, Lead, Body and Illustrations in Agricultural Information Items

Any information item when it is presented in the newspaper, generally has a heading, lead, body and illustration. It was observed that all the different typeś of agricultural information items had headings. Majority (63.42%) of the agricultural information items were without lead. It is essential to provide lead in agricultural information items also, since it introduces the subject to the reader and' expands the idea presented in the heading. Lead is possible when somebody is reporting some events, which goes on unfolding the story. This style may not hold good in reporting farm information e.g. 'cultivation of groundnut. Therefore, lead was absent in most of the farm information items.

Almost all (199.51%) agricultural information items had body. It was interesting to note that only two per cent. (1.95%) agricultural information items were illustrated. Illustration helps in getting the idea across the reader easily and quickly. It brings liveliness to the text and visualizes the contents. On an average, about 16 column cm. (0.38% to the total of space) space was occupied by the illustration in the agricultural information items. It shows most of the farm information items were without illustration and very little (0.02%) space was allotted to illustration. Probably, it may be due to unavailability of appropriate illustrations and its cost. As regards the contents of illustrations, these were appropriate to the text of agricultural information items. Black and white photographs were mostly used to illustrate farm information items in the newspaper studied. Though coloured illustrations are effective and attractive but seldom they were used.

5. Number of Words in Agricultural Information Items

Number of words in news determines size of news items. The data reveals that, majority (72 per cent) of the agricultural information items had upto -200 words, while 29 per cent farm informations items had more than 200 words. Most of the agricultural information items were in small size. This is because the newspaper contents were mostly urban oriented and necessarily more space was devoted to non-farm matters. Agricultural matter occupied very less space (2.32%) because, editors can't avoid farm information items totally, but they select and edit agricultural news and adjust it in a small space. It supports the findings reported by Mathur (1984), Sidhu et al. (1983).

6. Difficult Words in Agricultural Information Items

The percentage of difficult words in the text determines reading ease and comprehension (Flesch, 1960). The data regarding the percentage of difficult words in the agricultural information items reveals that large number (96.58%) of agricultural informations in the newspaper had more than six per cent difficult words. Slightly more than three per cent (3.4-2%) agricultural information items had less than six per cent difficult words.

A limit of difficult words should not exceed six per cent (Sinnarkar, 1966), otherwise difficulty in reading **Marathi** text increases, while reading ease and comprehension of the **Marathi** text decreases. As per the standard laid down by Sinnarkar (1966), it can be said the most of the agricultural information items appeared were difficult to read and understand. Farm information items published in the newspaper were not written by keeping in view the educational level of its farm and village readers. They may find it difficult to read and the purpose for which farm information items is published may be defeated. There is a need to present farm information items with minimum difficult words for the benefit of neoliterates and poorly literate village readers and farmers. Sinnarkar (1966) had recommended to use more number of simple words in the text meant for farmers so that low level literate can understand it. It is not possible to write on farming subjects by keeping the level of difficult words upto six or less than six per cent because farm writer may have faced difficulty in organizing and presenting technical facts. Most of the farm information items may not be directly useful to farmers, but it may be useful for readers other than farmers. That is why it might have been written in such a language that it contained more than six per cent difficult words.

7. Simple Words in Agricultural Information Items

Simple words in agricultural information items were studied. The use of simple words in English language is recommended by Flesch (1960) for increasing readability of text. In the present investigation only four per cent (3.9%) of the agricultural information items published in the newspaper had 50 per cent simple words. It was observed that, nearly 96 per cent (96.10%) published farm information items had more than 51 per cent simple words. This may be due to difficulty in presenting, organizing the technical matters on farming and lack of equivalent words used in local language and dialects for the technical terms or words used in farm writings. If vernacular words are used, it may be possible to use more number of simple words to increase readability of farm information.

8. Technical Words in Farm Information Items

Technical words in the agricultural information item were siudied. It was revealed that, about twenty five per cent (24.88%) farm information items published in the newspaper had less than one per cent technical words. About 48 per cent (47.80%) farm information items published in the newspaper had technical words between 1 to 6 per cent. And remaining one fourth (27.32%) farm information items published had more than six per cent technical words. Farm writers and editors had used technical words upto 3 per cent. It is difficult to write on farm information items without using technical words. There is a need to find out vernacular words which can be used conveniently in farm writings meant for farmers.

9. Personal Words in Farm Information Items

An attempt was made to study the per cent of personal words in agricultural information items. Use of personal words appeal to the readers and it make the writing lively and interesting. The reader Identifies himself with the situation described in news. If the percentage of personal words *are* less than six per cent it affects the readability of the text (Flesch, 1960). Scientific material of course, may be written without any personal words but it becomes difficult to understand. Yet, information on farming can be written by using personal words.

In the present investigation, only one fifth (21.46%) of the farm information items had 6 per cent or more personal words. This Indicates that the remaining farm information items had less personal words and may be considered as difficult to read. This may be due to either lack of training on the part of farm writers in reporting the farm Information items.

10. Length of Sentence in Agricultural Information Items

Length of a sentence is an important criterion of determining readability of text (Tayior, 1957). If the sentences are lengthy; low level neo-1iterate can't understand it. Therefore, short sentences are recommended while writing. It was observed that, about one fourth (25.8%) of the agricultural information items had 8 words per sentence. Easy prose is often written in 8 words per sentence or so and such writing is easy to read (Flesch, 1960). While standard of 17 words per sentence is maintained in only 17.07 per cent agricultural information items. It was observed that majority of the farm information items (79.81%) had sentences of less than 17 words. An average reader will have no difficulty while reading text having sentence; of 17 words (Kaur and Mathur, 1979). And about one fifth (20.9%) of the agricultural information items had more than 17 words per sentence. The difficulty in reading and understanding the text goes on increasing as the number of words per sentence exceeds 17 (Kaur and Mathur, 1979). It would be interesting to study these types of farm information items where a sentence had more than 17 words.

11. Number of Sentence in Agricultural Information Items

Number of sentences in the agricultural information items or news decides its size. It was revealed that about 76 per cent (76.10%) of the agricultural information items had 20 sentences and remaining 2 per cent had more than 20 sentences. The agricultural information items may have meaningful and readable text. The text preferably should be easy to read and comprehend for the low level literate. Lengthy agricultural information items result into low communicability and less effectiveness. In the present study, most of the farm information items published in newspaper were brief (20 sentences) .

12. Simple and Compound Sentences in Agricultural Information Items

Any publication should be written in simple sentences (Ensminger, 1957) for easy reading and comprehension. Number of simple sentences in farm information items appeared in newspaper was studied. It was found, that about 35 per cent (35.12%) of the agricultural information items had 50 per cent simple sentences and rest of the (65 per cent) agricultural information items had more than 50 per cent simple sentences. Flesch (1960) recommended the use of simple sentences in the text for easy reading. It was also revealed that about three fourth (77.87%) of the agricultural information items had upto 50 per cent compound sentences and remaining one forth had more than 50 per cent compound sentences. Compound sentences, particularly in the agricultural informations items, should be avoided, because low level literates find it difficult to read. They may not be able to read and understand compound sentences used in the agricultural information items.

13. Sentences in Heading, Lead and Body of Agricultural Information Items

Large majority (93.66%) of the agricultural information items had one sentence in the headings. The heading is at the top of news or feature article and it is first read. It conveys the essence of the information items and therefore, editors try to present heading in one sentence. Generally. the heading is supported by the lead. More than one third of the agricultural information items had lead in the present investigation.

Nearly twenty seven per cent (27.32%) agricultural information items had lead with one sentence while nine per cent (9.27%) agricultural information items had more than one sentence in leads. While rest of the agricultural information items were without lead, which may be due to the nature of the farm article and type of information. More than fifty per cent of the agricultural information items had 1 to 10 sentences in the body and remaining had more than ten sentences in the body. It again indicates that agricultural information items published had short text.

Conclusions

Farm information items appeared in **Sakal** a leading newspaper - were analyzed for its certain textual characteristics. The agricultural information items occupied meagre space in the **Sakal** - a leading newspaper from Maharashtra - appeared on right hand side of the middle pages and rarely of first and last page. The heading of agricultural information items was mostly flush left and crossline type. Farm information items had heading, body but had no illustration and lead. Photograph was the most commonly used in the farm illustrations. The size of farm information items was of 200 words. Most of the farm information items had more than six per cent difficult words, more than 70 per cent simple words, 1 to 6 per cent technical words and less than two per cent personal words. The average length of sentence ranged from 8 to 17 sentences. Majority of the farm information items had more than 50 per cent simple sentences and 50 per cent compound sentences. Most of the farm information items had one sentence in the heading and the lead, while 10 sentences in the body.

REFERENCES

1. Crump, Spencer., "Fundamentals of Journalism" Mc-Graw-Hill Book Company, Inc. New York, P.159, 1974.

2. Ensminger, O., "A Guide to Community Development', Ministry of Communicty Development, Government of India, New Delhi, 1937.
3. Flesch, R., "How to Write, Speak and Think More Effectively", A Signet Book, New American Library, Times Mirror, New York, P.10-11, 1960.
4. Holtje, H.F., "Theory and Problems of Advertising", Me Graw-Hill Book Company, New York, 1978.
5. Joshi, D.L. and M,G.Kulkarni., "The Role of Mass Media in Rural areas", Marathwada University Journal, 3:17 & 18:23, 1978-79.
6. Kaur, O.K. and P.N.Mathur., "Effective Feature Writinq for Farmer Readers", Indian Journal of Extension Education, 15:3 & 4:27-33, 1979.
7. Mathur, I.K., "Agricultural Journalism 50 years : A Case Study of Papers From Hyderabad", A paper presented for the National Seminar on Agricultural Journalism IM'jblems and Prospects, GBPA&T, Pantnagar, 27th to 29th March,'84, 1984.
8. Sinnarkar, N.P., "A Suggested Score fnr Rating Readability of Extension Literature in M a < a t h i", Agriculture College Parbhani, Magazine, 9:6^-68, 1966.
9. Sidhu, O.K., K.C.Tyagi and Mishra, "Content and Coverage of Agriculture and Dairy News in Regional & National Dailies" Interaction, 1:1:77-90, 1983.
10. Sawant, R. K., and S.D. Waghmode, "Reading Behaviour of litrate Villaqers of $r_3$gha Village in Ahmadnagar District". The Maharashtra Journal of Extension Education 4:1 s1 43-1 44, 1985.
11. Taylor, W. L., "Cloze Readability Scoies on Indices of Individual Differences in C ornpr enension and Attitude" Journal of Applied Psychology, 4:19:26, 1957.

Chapter **26**

Utilization Pattern of the Farm Publication's

Rajendra Rathore, R. N. Trikha, Anita Singh

Mass media plays a significant role in developing rural resources and motivating the masses for the adoption of new technologies. The mass media is a very powerful instrument which if properly used, can cause unprecedented social change and knowledge advancement. Information through farm publications are perhaps the most favoured and widely adopted among the farmer's community. There is every likelihood of enhanced reading and proper utilization of the information.

The Rajasthan Agricultural University Bikaner has been bringing out the farm publication regularly for more than last two decades in the area of Agriculture, Animal Science, Agricultural Engineering & H. Sc. The purpose behind the production of such literature is to motivate the people for the desired action and for the adoption of new technologies. Therefore, the present study was undertaken with the following specific objectives.

1. To study the socio-personal characteristics and communication profile of the respondents.
2. To determine the utilization pattern of the farm publications.
3. To study the correlation of socio personal variables with the utilization pattern of the farm publication.

Methodology :

The study was purposively conducted in the Rajasthan State the study was restricted to six district out of 32 districts each having one K.V.K. The district having higher member of subscribers from each administration division were selected for the study. The subscribers of **"Apna Patra"** during 1998 were over 3500 in the state. The member of subscribers differ in each district. Therefore, 20 percent sample of the respondent from each sample district was selected through simple random technique. Thus, a sample of 256 respondents constituted the sample. Other extension publications (Folders & Booklets) published during the year 1998 was taken for the study. The exploratory research desing was used to assess the readers profile. Utilization pattern and relationship of some independent variables with the utilization pattern of the publication.

The data were collected through structured interview schedule, personal contact, discussion and observation. The data were analyzed using simple statistical techniques. Frequency percentage, mean weighted score, S. D. and correlation co-efficient.

Table-1. General profile of the respondents :

S. No.	Variable	Respondents (N = 256)	
		Frequency	%
1.	Age		
	Young	109	42.57
	Young	125	48.83
	Old	22	8.60
2.	Education		
	Illiterate	08	3.12
	Primary	130	50.78
	Middle	50	19.53
	High School	49	19.14
	Graduate and above	19	7.42
3.	Caste		
	Schedule caste/Schedule tribe	58	22.66
	Backward caste	53	20.70
	Others	145	56.64
4.	Occupation		
	Farming	197	76.95
	Business	09	3.52
	Service	07	2.73
	Farming with business	22	8.60
5.	Land holding		
	Marginal (< 1 ha.)	29	11.33
	Small (1 to 2 ha.)	58	22.666
	Medium (2 to 5 ha.)	80	31.25
	Large (>5 ha.)	89	34.76
6.	Extension contact		
	Low	50	19.53
	Medium	176	68.75
	High	30	11.72

S. No.	Variable	Respondents (N = 256)	
		Frequency	%
7.	Media exposure		
	Low	62	24.22
	Medium	171	66.80
	High	23	8.98
8.	Social participation		
	No membership	156	60.94
	One membership	90	35.16
	More than one	10	3.91

From table-1, it is clear that majority of the respondents were of 'middle' age, educated up to primary level belonging to the other castes, having farming as their main occupation and had large land holdings. As far as mass media exposure extension contact degree of social participation were concerned, it was found that majority had medium level of mass media exposure, extension contact and no membership of any organization.

Table-2. Distribution of the respondents according to the reasons for subscribing the farm publications :

Reasons	Frequency	Percentage
Own reading	250	97.66
Gram panchayat	04	1.56
Research purpose	02	0.78

The data in the table-2 show that majority of the respondents (97.66 percent) subscribe to the publication for their own reading followed by community reading in Gram Panchayats (1.56 percent) and for research purposes (0.78 percent).

Table-3. Distribution of the respondents according to the reasons for the farm publications :

Reasons for reading farm publications	Frequency	Percentage
Increase knowledge	132	51.56
Exchange information	97	37.39
Encourage others	55	21.48
Awareness	68	26.56
Adoption	107	65.23
Innovation	16	6.25

Table-3 indicates that the majority of the respondents (65.23 percent) reported reading farm publications for adoption of practices in their own field as one of the reasons followed by the increase in knowledge (51.56 percent), exchange of information (37.89 percent). Awareness about the new information and ideas (26.56 percent), and encouraging other (21.48 percent).

Only 6.25 percent respondents reported innovation as a reason.

Table-4. Distribution of the respondents according to the clarification sought after reading the farm publications:

Behaviour after reading the publication	Frequency	Percentage
Discussion with family members	150	58.59
Discussion with others	146	57.03
Maintain a diary	64	25.00
Seeking additional information	54	21.09

Table-4 reveals that the majority of the respondents (58.59 percent) have a discussion with family members after reading the farm publications followed by the discussion with others (57.03 percent), maintenance of diary (25.00 percent). Only 21.09 percent respondents seek additional information after reading.

Table-5. Distribution of the respondents according to the method of query :

Methods of query	Frequency	Percentage
Letter writing	30	11.72
Self approach at centre	203	79.30
Clarification with experts on their visit	130	50.78

The data in table-5 shows that the self approach at the centre for clarification was maximum (79.30 percent) followed by the clarification during visit of experts (50.78 percent) and letter writing (11.72 percent).

Table-6. Distribution of the respondents according to the reading pattern :

Reading patterns	Frequency	Percentage
Complete	187	73.05
Partial	61	23.83
Not at all	8	3.12

The table-6 shows that the majority of the respondents (73.05 percent) read the farm publications completely followed by the partial reading (23.83 percent). Only 3.12 percent respondents did not read the publications at all.

Table-7. Distribution of the respondents according to the time spent for reading :

Time spent for reading	Frequency	Percentage
Up to an hour	67	26.17
1-2 hours	93	36.33
More than 2 hours	96	37.50

The data in table-7 reveals that majority of the respondents (37.50 percent) devoted more than two hours for reading the farm publications followed by the one to two hours (36.33 percent) while 26.17 percent devoted up to an hour. Almost similar trend was identified by Zalaki (1973).

Table-8. Distribution of the respondents according to the application of the agricultural information :

Application of agricultural information	Frequency	Percentage
Selection of new varieties crops	50	19.53
Hybrid seeds	43	16.80
Fertilizer use	49	19.14
Increase in yield and profit	119	46.48
Pesticides use	16	6.25
Weed control	11	4.30
Land fertility	7	2.73
Other practices	82	32.03

Table-8 reveals that majority of the respondents (46.48 percent) reported the utilization of agricultural information for increase in yield and profit from the crops followed by other practices (32.03 percent), selection of new variety of crops (19.53 percent), fertilizers (19.14 percent), hybrid seeds (16.80 percent), new pesticides (6.25 percent), and weed control (4.30 percent). Only 2.73 percent respondents reported the utilization of the information for the land fertility.

Table-9. Distribution of the respondents according to the preservation of publications after reading :

Presentation	Frequency	Percentage
Filling articles	39	15.23
Filling of recommendations	36	14.06
Keeping copies	171	66.80
Binding of publications	29	1133
Throw as waste	7	2.73

Table-9 indicates that majority of the respondents (66.80 percent) preserve the farm publication for further use by keeping the copies safely, followed by filling of articles (15.23 percent), filling of recommendations (14.06 percent), binding of the publications (11.33 percent), and throwing as waste (2.73 percent).

Table-10. Distribution of the respondents according to the utilization of the information :

Utilization of the information	Frequency	Percentage
As such	44	17.19
After modification	200	78.13
After discussion with progressive farmers	77	30.08

Table-10 reveals that majority of the respondents (78.13 percent) were utilizing the information after modifications followed by the discussions with progressive farmers (30.08 percent), and using the information as such (17.19 percent).

Table-11. Correlation coefficient between independent variables utilization pattern :

S. No.	Variables	'r' values
1.	Age	0.289**
2.	Education	0.1475*
3.	Caste	0.1851*
4.	Occupation	-0.1855*
5.	Land	-0.0259 NS
6.	Mass media exposure	0.897 NS
7.	Extension contact	0.1810*
8.	Social participation	0.3238**

*Significant at 1 percent level

A perusal of data in the table-11 reveal that age, education, cast, extension contact, social participation were positively and significantly associated with the utilization pattern of farm publications at 1 percent level of probability. While occupation was negatively and significantly related with the utilization pattern of farm publication, it means that these six variables (age, education, caste, occupation, extension contact and social participation) exerted their influence on the utilization of farm publications published by the RAU. However, remaining independent variables viz., land and mass media exposure level were found in non-significant association with utilization of farm publications by the farmers. It means that these variables did not exert any significant impart on the utilization pattern.

Conclusion :

The findings of the study indicates that the subscribers of farm publications were form different agro-ecological and socio-economic groups of Rajashthan. Most of respond were middle aged, educated up to primary level, having large land holdings and medium level of extension contact.

The utilization of farm publications was at the medium level, however, most of the farmers utilized the information after modification and seeking clarification of information from research centre.

Age education, caste, extension contact and social participation were positively significant in farms of utilization pattern.

Reference :

1. Chidanandappa, G. and Veerabhadriah, U. 1988. Mass media utilization by extension personnel. Mysore J. Ag. Sci. 22 (1) : 120-125.
2. Karande, A. G. and Riswadkar, M. R. 1987. Use of print media in information communication in animal husbandry. Maharashtra J. ext. Edu. 6(4) : 51-54.
3. Mathur, P. N. Singh, K. N. and Das, K. K. 1978. Readership patterns nature of content, sources of information of farm periodicals in India. Indian J. Ext. Edu. 14 (3&4) : 39-44.
4. Nataraju, M. S. 1996. Improving effectiveness of farm periodicals. IASSI Quart. 15(1) : 158-165.
5. Natikar, K. V. Govimath, U. S. and Budihal, R. A. 1995. Reading behaviour of the farmers of northern Karnataka. Indian J. Ext. Educ. 21(1&4) : 89-90.

6. Pandey P. and Trikha, R. N. 1990. Effectiveness of home science extension folder as perceived by rural women. Indian J. Ext. Educ 26 (1&2) : 115-117.
7. Saha S. and Trikha. R. N. 1989. A study on contents and reader's characteristics of Indian Farmers' Digest. Indian J. Ext. Edu. 25(1&2) : 58-60.
8. Udomissor, I. W. 1997. Media influence and media habit of Indians and Nigerias. Communicator. 32 (11) : 37-44.

Chapter 27

Print Media and Agricultural Development

Dr. Ranjit Singh

The modern farm technology evolved through a systematic agricultural research has tremendously helped the farmers to increase farm production and productivity. At the same time, it has also been established that only a part of the new technology has reached the farmers. There is lot of scope for increase in production because the level of yield at farmers' fields is far below the demonstrated results. Efforts should be made to bridge the wide knowledge arid productiongap. The success in farming now depends upon the adoption of scientific technology. It is, therefore, essential that farm information must be disseminated in such a way that it reaches the farmers quickly, accurately and timely. Experimental findings have clearly proved that mass media, if properly utilized can influence the farmers in the adoption of new farm technology. Thus, it can play a vital role in reduction of technological gap and will help in bringing a significant increase in the farm production and productivity. Print media is one of the important media of mass communication along with films, radio and television.

Print Media

It is believed that print media demands a more active creative participation on the part of the reader than is demanded by the audiences of other media because the communicationis less structured and reader is forced to participate creatively in this more impersonal type of communication. Such creative participation is supposed to be persuasively advantageous. According to Klapper (1966), print alone among the media, allows the reader to control the occasion, the pace and the direction of his exposure and permits him easy re-exposure, more easily than other media, print allows a topic to be developed to whatever length and with whatever complexity seems desirable. Goldstein (1940) suggests that print produces superior retention of complex factual material than does oral presentation. As compared to other media, print can be readily used to reach small and specialised audiences for which other media would be prohibitively expensive. It is traditionally associated with culture and may carry a higher prestige for some people than do the other media. The printed word has a lasting-power far beyond that of the spoken word or the visual; image. Readers can refer to it again and again.

Print Media and Agricultural Development

The role of print media in agricultural communication-needs to be emphasised. A question is often posed on the wisdom of spending money and energy on publications when most of our farmers cannot read and write. Here, one must be clear that illiteracy is a relative term as absolute illiteracy in the community hardly exists. A community, even with a few literates can spread the message conveyed through the printed words.

Farm literature can accelerate the process of development and can guide it into channels that ensure the maximum benefit to these layers of rural society which need-socio-economic upliftment the most. Therefore, farm literature is not merely to propagate or disseminate the simple mechanics of farm technology but has to take in all factors-concerning the process of rural development. The main-components of print media can be newspapers, magazines, leaflets and bulletins. Since the newspapers have a large and varied readership, they also possess the potentialities of communicating farm information to a large number of people simultaneously and on quick pace, The newpapers are also a low cost medium of communication. The increasing literacy rate in the countryside offers new promises and prospects for the utilization of newspapers as a medium of mass education. In the modern context, when everybody is interested in a quicker development of agriculture as a; booster of economy, the social responsibility of the press in terms of creating such an awareness in the rural sector cannot be over-emphasised. The newspapers constitute an important medium for the transfer of technology from the centers of origin to the ultimate consumers. The newspapers serve the people in three ways i.e. inform, guide and entertain. They can provide farmers with precise and digestible information regarding the use of fertilizers, pesticides, insecticides as well as with the information about the latest' recommendations on various farm aspects. For most people the newspaper has a more vivid personality than any of the-other media. The disappearance of newspapers from a city creates confusion in business life and in the ordinary flow of social and civic affairs. The subscriber without his newspapers feels as lonesome as a person whose best friend has left the town. This strong appeal of the press can best be utilized for transfer of technology. Once farmers become addict to the habit of reading newspapers it will become a permanent media to talk to the farmers.

It is a myth that print media is not capable of transferring technology. According to this belief scientific information is difficult to transfer through distance education without practical demonstration. This myth has already been exploded by different research studies where it has been established that it is possible to transfer technology effectively and efficiently through print medium. To further substantiate it, I would like to cite two of my personal experiences : (i) Some Indian Universities started correspondence courses -in sixties. An idea came to my mind to make use of this system for providing scientific knowledge to the farmers. I was of the firm belief that to get the maximum advantages from the new technology and to practice scientific farming. One must have a systematic and complete exposure to the knowledge of scientific agriculture. Credibility of scientific methods of farming and of written words have already been established. Moreover, in Punjab majority of the farming families have atleast one educated member. Written material has the advantage that learning occurs when and where it is convenient to the learner and at the rate of assimilation set by the individual. Moreover, literature is usually well designed and professionally produced. Inspite of the strong opposition, this project was started as an experiment in 1972. This correspondence course came as a boon to the busy farmsrs of Punjab who could not remain away from their farms and homes for a long period to participate in training courses and young farmers who have decided to adopt farming as profession. The success of this course prompted to start another course for farm ladies in 1973. Another course for small and marginal farmers was started in 1976. Sincere efforts were made to prepare the booklets. Contents were identified on the basis of information needs of the different categories of farmers. Lessons were got written from the Subject-Matter Specialists and got edited by the Language and Communication experts so that the contents: are technically accurate, specific and presentable and at the level of understanding of the farmers. In order to make sure that the lessons have been read and comprehended, a question paper is attached with each lesson. The trainees were required to answer the same along with difficulties, if any. Based on the responses of the trainees and inflow of new technology, the booklets are revised after every three years.

Evaluation of studies conducted regarding the usefulness of the correspondence courses has brought out that print media is capable of transferring technology effectively, timely and promptly. Saini (1979) studied the responses of the small farmers who have completed the one year course. According to him, the distance teaching was efficient in the fulfilment of information needs of the trainees in respect of agricultural technology. The majority of the respondents could grasp reasonably the contents of the lessons. The lessons proved useful to the respondents not only in informing, but also ensuring the application in the field situation of information related to the improved practices of the selected crops.

Similarly, Dhillon (1978) found that the majority of the farmers could comprehend the contents of the selected lessons. Most of the farmers perceived that the information given in the lessons was highly relevant to their needs. The majority of the farmers used the information obtained, from lessons in their field situations.

As far as the farm women are concerned, Singh and : Saini (1983) obtained the similar results.

I am of the firm belief that newspapers specially the vernacular press can play a very significant role in agriculture and rural development. When I took over the charge of the Communication Centre, I persuaded the Editors of Punjabi newspapers in Punjab to start agricultural supplements also on the pattern of supplements on sports, films, women, etc. Most of the Editors agreed to start fortnightly one-page supplement. Although no formal study has been conducted to assess the impact of these supplements, but the letters being received by the Editors, gave an indication of their popularity. These supplements have increased the popularity of the newspapers in the villages and this is one of the reasons for a considerable increase in their circulation. Impressed by the popularity of these supplements two leading Punjabi dailies 'Ajit' and 'Punjabi Tribune' have increased the frequency from fortnightly to weekly supplements.

These experiences and the findings of research studies **have** confirmed that print medium or distance education has great potential in the communication process for quick transfer of technology.

Pre-Requisites

Moreover, there appears to be certain peculiar prerequisites that determine the possibility of the success of this media in the transfer of technology :

1. Availability of sound technology that is superior to the prevalent practices.
2. A realization among the farming community that superior technology exists and they can profit by acquiring it.
3. Qualified and trained communicators who can successfully transmit the message to readers so that they can grasp it and utilize it at their respective farms.
4. A desire on the part of the farmers to acquire new technology and their faith in scientific technology to solve their day-to-day problems.
5. There are increasing numbers of problems that need quick solutions and can be solved by the specialists.
6. The channel is more useful to the farmers who take initiative to obtain information individually at their own. It is not the laggards who will be the potential readers of literature.
7. Utility of the print media will be governed by the level of literacy prevailing in the farming community and also the stage of agricultural development.
8. This media is more popular with the younger generation of farmers.

History and Present Status :

It is believed that Portaguese Missionaries were the first to set up printing press in India at Goa in 1552 (Goel, 1980). The book entitled "Catechsimode Dectrine" was printed in this press in 1956. Almost all the big cities got printing presses by the end of eighteenth century. 'Bombay Carrier' was the first newspaper started by Rustomji Keyshapji in 1722. The credit for printing in Indian scripts goes to Silliam Carrey of Serirampore of West Bengal.

Bengal Famine forced the British Government in India to make efforts for the development of agriculture in India. As a part of the efforts to develop agriculture 'Bengal Agricultural Gazette' under the editorship of Polart Knight -was the first periodical on agriculture. Another periodical was started towards the end of 19th century entitled 'Plantation News' mainly for the European planters. In fact, 'Indian Farming' was the first regular magazine on farming and was started in 1930. At present about 200 farm and home periodicals are produced in various languages in the country. Majority of these are in Hindi (37%) and English (20%) and are owned (20%) by Goverment and semi-Government organizations. Most of the Editors of these papers are without rural background and training in journalism.

According to Singh (1984), a publication is called a rural one if it devotes a minimum of 40 per cent of its space meant for communication to rural coverage. The anatomy of a privately managed newspaper or of a magazine is bound to keep its urban base, because the major source of its income comes from advertisements which mostly cater to the urban readers and cover sixty per cent of the total space for the sheer economic survival of the periodical. Hence, it is a misnomer to call a privately managed periodical a farm publication inspite of its having been categorised as a rural one, because it devotes hardly 16 per cent of its total space to rural coverage.

Most of the farm magazines and bulletins are being-published by the agricultural universities. Central Farm Information Bureau which was established in 1958 in the Directorate of Extension, Ministry of Agriculture also communicate farm technology through newsletters, leaflets, pamphlets and bulletins. The literature produced by agricultural universities in regional language has done a commendable job in communicating the farm technology. For example, the Punjab Agicultural University brings outmonthly magazines both in Punjabi and English, *Changi Kheti* and *Progressive Farming.* Though the contributors in both the periodicals are University teachers who carry on their official work of teaching and research in English, yet the Editor of *Changi Kheti* tries to ensure that it does not become a satellite of *Progressive Farming.* It goes to the credit of *Changi Kheti* that it tops in circulation among monthlies m Punjabi and has a proud record of contribution to the rural development in Punjab. One of the most significant contributions of the Communication Centre of Punjab Agricultural University is to inculcate the habit of reading among the rural people. The literture produced by the Centre hasbecome a craze for the farmers only next to the seeds of high-yielding varieties. This Centre has produced 432 bulletins (66 in 1968) on different aspects of farming. The annual sale of literature which is moderately priced has touched Rs. 5 lakh in 1984 from Rs. 50, 000 in 1968. During the Kisan

Mela in March 1985, the two-day sale of literature was about Rs. 42,000/-. The circulation of *Changi Kheti* which was 1, 500 in 1968 touched 12,000 in 1984. All the publications are financed through a Revolving Fund without any financial aid or subsidy.

Even after 38 years of independence, Indian press has given low priority to agriculture A few English dailies namely 'Times of India' and 'Hindu' have been releasing special farm features. As they have not been able to devote more than five per cent of their news-space or to say precisely two per cent of their total space to developmental journalism and their contributors could at the most write about rural areas and not for rural areas, their exercises have been missing the real target. Such farm journalists due to their ignorance of real atmosphere and conditions have been creating myths about the rural prosperity, thus doing a dis-service to the farmers. According to Dosanj (1980) most of the newspapers are interested in agricultural news of the sort which either affect the price of a fresh milk, wheat, flour, pulses and other necessities or the birth of a freak calf, etc. By giving space to such news the urban based journalists not only play up sensationalism, but also confirm the prevalent city notion that agriculture cannot make a real news. These persons are ignorant or tend to become ignorant of the massive research which has made the Indian agriculture dynamic and the problems being faced by the cultivators. In fact, at present agricultural reporting occupies the lowest rung, in the reporters' ladder to prestige. As stated by Singh (1978) farm news and stories do find a place in the national press, but more of the newspapers use these on lean days. These are regarded as casual fillers.

Sodhi (1982) while studying the space devoted by two leading English dailies of Northern India namely 'Indian Express' and 'The Tribune' found that the Indian Express devoted 2.60% of the total space to the farm information in terms of area in square centimetres. The same information when calculated in terms of number of column lines was found to be 2.3 per cent. The total space occupied by farm information in The Tribune, in terms of area in square centimetres was 2.08 per cent. As much as 1.90 per cent of the coverage was covered in terms of number of column lines. In both the newspapers, the content area of agricultural marketing occupied the maximum coverage.

In another study, Dhillon (1982) while analysing the agricultural contents of three major Punjabi dailies, found that very little space was occupied by the various content areas relating to farm information in all the newspapers, viz. *A jit, Jagbani,* and *Punjabi Tribune.* The study disclosed that *Ajit, Jagbani* and *Punjabi Tribune* devoted only LOS per cent, 1.20 per cent and 2.74 per cent of the space to farm information, respectively.

The availability of newspapers and magazines in villages is also very low. According to Singh (1978), hardly two per cent of the papers reach villages where 70 per cent of population lives. To create minimum desirable awareness in a developing country, the UNESCO has laid down that 100 copies of newspapers for every one thousand people should be provided. We have only 13 newspapers for every one thousand people in our country. It remains a bitter truth that Indian press lacks a living contact especially with people living in villages.

Suggestions

It is an admitted fact now that print media can play a significant role in transfer of modern technology. It is also a fact that the contribution of print media so far is far below the expectations. On the basis of the discussion of this paper, following suggestions are put forward to make print media more useful. This is a challenge to farm communicators in the coming years.

1. Massage should be organized in such a way as to give package approach to the programmes with emphasis and motivation for adoption by farmers.
2. The type of journalists employed, mainly for political reporting are unlikely to meet the task. The challenge, therefore, is to carefully recruit and offer relevant training to an army of youngesters for this purpose.
3. To enhance the communicability of the programmes, there is a greater need to support it with the receiver's research and feed-back system. At present, it is almost a one-way traffic as there is little feedback.
4. Print media has so far been mainly used to create awareness, but there is a need to give complete information to facilitate adoption.
5. Most of the time, print media has been used as a tool of publicity. Communicators have to make a commitment to use it for development i.e. for education and information.
6. Print media is considered as a passive medium, but efforts have to be made to convert it into participatory medium, by encouraging readers' responses.

7. In order to sustain interest of rural people, scripts, should be imaginative, interesting and attention holding.
8. The communicators will have to be technically sound in addition to specialization in communication with special reference to print media.
9. The communicators have to show complete identification with the community which they seek to serve. They must involve themselves fully with every kind of activity that goes on in their area. They should feed information of all kinds which is now either wholly desired or relatively disseminated to the rural people.
10. The Directorates of Extension Education or Communication Centers of the Agricultural Universities may hold refresher courses in writing for the editorial staff and reporters of the periodicals to impart them training in the requisites stated above.
11. Writing for farm publications also includes translation of technical messages into regional language. Many translators who do not have the knowledge of local conditions create ludicrous versions, which are both erroneous and misleading. There is a need to prepare dictionaries of technical terms in regional languages.
12. At present, only those newspapers/periodicals are entitled to get advertisement which has a minimum circulation of 2000. This figure is lowered to 500 for Sanskrit papers and the ones meant for border and tribal areas. The farm magazines should receive the same consideration as extended to Sanskrit periodicals. This will encourage having farm magazines in regional languages relating to different agro-climatic regions.
13. Farm Journalism is a missionary job, which has an obligation to tell and does not resort to any manipulation to sell through scoops. It avoids the underworld, but champions the underdog. The writers, publishers and editors of farm periodicals have to keep this in mind in order to avoid frustrations.
14. The press should feel social responsibility towards two-third of the population and devote at least 10 per cent space for such coverage. Vernacular press should follow the example of Punjabi dailies of Punjab and start regular rural and farm supplements.
15. Most of the newspapers are interested in sensational news even in agriculture. But, their major emphasis should be to transfer

latest technology and to highlight the problems being faced by cultivators including small and marginal farmers and farm labour.

16. Farm literature does not mean merely propagating or disseminating the simple mechanics of farm technology, but must take in all factors concerning the process of rural development.

As the modern farming is becoming complex everyday, the print media can play a significant role in the transfer of modern technology. Though the majority of the readers tend to be rural elites, there is growing evidence that farmers from various socio-economic strata express keen interest in the farm journals and literature and influenced by it. At the same time, it is important to demonstrate that rural newspapers can be successful in not merely meeting an unsatisfied need, but can be sustained economically viable propositions.

References

1. Dhillon, D. S.: 1978 : "An appraïsal of the Correspondence Course Programme for the farmers" run by PAU. *M. Sc. Thesis, Punjab Agricultural University, Ludhiana.*
2. Dhillon, B.;1982 : "Content Analysis of farm information contained in selected Punjabi Dailies circulating in Punjab," *M. Sc. Thesis, Punjab Agricultural University, Ludhiana.*
3. Dosanj. S. S.; 1980 ; "Journalism as a channel of change and rural development", *Paper presented at the National Seminar on "Absorption of Development messages by Target Groups" held at National Institute of Rural Development, Hyderabad.*
4. **Goel,** A. C.**;** 1980 : "Theory and Practice of Composition", *Manimajra, Cole's Press.*
5. Goldstein, H.; 1940 : "Reading and Listening Comprehension at various controlled rates, *N. Y. Columbia University.*
6. **Klapper, J. T.;** 1966 : The effects of Mass Communication' *The Free Press, N. Y. P. 110.*
7. Saini, G. S.; 1970 : "Potentials of distance teaching—A Study of the correspondence courses for small and marginal farmers of Punjab state.," *Ph. D. Thesis, Punjab Agricultural University, Ludhiana.*
8. Singh, P. P.; 1978 : "Role of Farm Journalism in Rural Development," *M. Sc. Thesis, Punjab Agricultural University, Ludhiana, p. 13.*
9. **Singh, R.** and Saini, R. K.; 1983 : "Opinion of trainees about the

correspondence course for farm ladies. *J. of Res. 203. pp : 353-58.*

10. Singh, H ; 1984 : "Problems of editing farm magazines." *Advance Dec. pp. 31-34.*

11. Sodhi, S.; 1982: "Content analysis of farm information contained in selected English Dailies circulating in Punjab." M. Sc. *Thesis, Punjab Agricultural University Ludhiaria.*

Chapter **28**

Information Sources Availability and Utilization Pattern Among Farm Families

Upayana Singh, O.P. Mishra[2]

Mass media like T.V., Radio, Newspaper and Magazines are playing a key role in information dissemination to rural people in general and farming community in particular. Information dissemination strategy can be made more effective if information sources availability and utilization pattern among the farm families is known to the media managers and planners. Taking this backdrop, the present study was conducted with the objective to know the availability and utilization pattern of information sources among farm families.

Research Methodology

The present study was conducted in the two purposively selected villages (viz. Bhagwanpur and Ramnagar) of Rudrapur Block of District U.S. Nagar (Uttaranchal). The data were collected during January to March 2002. All the 105 farm families from these two villages were taken as the respondents for the study. The head men and head women of each family were interviewed. Personal interview technique was used to collect data on specially structured interview schedule for the study. The data collected were analyzed by applying simple statistical technique like frequency and percentage.

Results and Discussions :

Availability of Information Sources

Availability of mass-media, such as television (T.V.), radio, newspaper and magazine among the respondents was studied. Data related to availability of mass media has been presented in Table-1.

Table-1 : Type of media available.

N=105

S. No.	Media	Available		Not available	
		Frequency	Percentage	Frequency	Percentage
1.	TV	94	89.52	11	10.48
2.	Radio	60	57.14	45	42.86
3.	Newspaper	32	30.48	73	60.52
4.	Magazine	32	30.48	73	69.52

It is evident from Table-1 that a maximum number of respondents (89.52%) had T.V. sets with them. About 57 per cent of respondents had radio sets. Newspaper and magazine subscribers were to the extent of 30.48 per cent each. However, only 5.71% of the respondents had no media available with them. Although, several researches have indicated that radio was available to majority of the respondents followed by T.V. (Kukreti, 1990; Sangha and Kalra, 1993; Pandey, 1997; Begum, 2000; Kumar 2000) but the results of the present study do not seem to be in line of above findings. In this study, it has been found that majority of the respondents possess T.V. followed by Radio.

As regard to the availability of newspaper and magazine, the results of the present study are similar to the findings of the Sangha and Kalra (1993). Singh (1997), Nanjappa and Ganapathy (1986).

Number of Mass Media Owned by the Respondents

It is evident from the data presented in Table-2 that maximum number of respondents (35.24%) had two media available with them followed by one medium (29.52%), four media (19.04%) and three media (10.48%).

Table-2 : Number of media subscribed (owned) by the respondents.

N = 105

Category	Frequency	Percentage
Persons having one media	**31**	**29.52**
Persons having two media	**37**	**35.24**
Persons having three media	**11**	**10.48**
Persons having four media	**20**	**19.05**
Persons without any media	**6**	**5.71**

Combination wise Availability of Mass Media

Data presented in Table-3 reveals that out of total number of respondents, 24.76 per cent of respondents had T.V. as the single medium available with them. An equal number of respondents (24.76%) had the combination of T.V. and Radio followed by T.V., Radio, Newspaper and magazine (19.04%). A combination of T.V. and newspaper was owned by 6.66% of respondents. It was seen that 5.71% of respondents had combination of Radio, T.V. and magazine while 4.76% of respondents owned radio as the only source of media. T.V. and magazines were available to 3.80% of respondents followed by 2.86% of respondents who were having T.V., radio and newspaper. However, only 1.90% of respondents were having T.V., newspaper and magazine as a combination available.

Table-3 Availability of mass media according to various combinations.

N=105

S. No.	Combinations of media	Frequency	Percentage
1.	TV	26	24.76
2.	TV and radio	26	24.76
3.	TV, radio, newspaper and magazine	20	19.04
4.	TV and newspaper	7	6.66
5.	Radio, TV, magazine	6	5.71
6.	No media	6	5.71
7.	Radio	5	4.76
8.	TV and magazine	4	3.80
9.	TV, radio and newspaper	3	2.86
10.	TV, newspaper and magazine	2	1.90

Utilization Pattern of Information Sources

The extent and pattern of mass media utilization including type of media utilized, media utilization of their own or of others, rate of utilization, purpose of utilization, segment of the day when mass media utilized, amount of time spent with mass media and utilization of different combinations of mass media were studied. The results of all the above aspects have been discussed below.

Type of Media Utilized By The Respondents

T.V. was found to be Maximum Utilized Mass-Media among all the other media. About 99% Of Respondents Utilized T.V. Followed By Newspaper (85.71%). Radio And Magazines Were Equally Utilized By The Respondents I.E. 53.33% Respondents each (Table 4).

Table-4 : Mass media used by the farm families

N=105

Mass media	Yes		No	
	Frequency	Percentage	Frequency	Percentage
T.V.	104	99.05	1	0.95
Radio	56	53.33	49	46.66
Newspaper	90	85.71	15	14.28
Magazines	56	53.33	49	46.66

It seems from the result that T.V. is most popular and most utilized source of information. Even people who are not possessing T.V. sets used to watch T.V. programmes in others houses. As T.V. is an audio-visual aid, it gives pleasure to eye and ear of the listener. It gives pleasure to note that newspapers are also utilized by large percentage of farm families. If we see the availability data, it was being subscribed by about 30% of the respondents whereas it was being utilized by about 86% of the respondents. This shows that newspaper has got high degree of popularity among the respondents. Percentage of radio owners in the selected villages were more as compared to radio users. It makes clear that even people having radio sets were not utilizing it, It may be due to availability of T.V. sets in the house. People seems to give more preference to T.V. as compared to Radio as a means of information. Another reason may be that people in the "Tarai" area of Uttaranchal state are economically sound and availability of electricity is also high as compared to many other parts of the country. These two conditions are favourable for more utilization of T.V. than radio.

Findings of the present study partially support the findings of the Kumar (2000). However, findings of the present study are not in line with findings of Chauhan and Singh (2001).

Media Utilization of their own or of Others

Television : Maximum number of respondents (86.54%) were using their own T.V. sets. Only 13.46% of respondents were using T.V. sets of others (Table 5).

Radio : Out of total radio users, 80.36% of respondents used their own radio sets where as 19.64% of respondents used radio sets of others.

Newspaper : It was found that a majority of respondent (4.44%) out of total newspaper readers used newspaper of other. The

Table-5 : Mass media used of their own or others.

Mass media	Own		Others	
	Frequency	Percentage	Frequency	Percentage
TV N = 104	90	86.54	14	13.46
Radio N = 56	45	80.36	11	19.64
Newspaper N = 90	32	35.56	58	64.66
Magazines N = 56	31	55.36	25	44.64

remaining 35.56% of respondents read their own newspaper.

Magazines : Out of the total magazine, readers 55.36% of respondents used their own magazines whereas 44.64% of respondents read magazines of others.

RATE OF UTILIZATION

The rate or extent of utilizationof each media by the respondents was studied in terms of regular, some times and never used. The result has been presented in Table-6 and have been discussed below.

Television : Out of total number of T.V. users 83.81% of respondents were regular viewers followed by casual viewers (15.24%).

Table-6 : Rate of mass media utilization.

N= 105

Mass media	Regular		Sometimes		Never	
	Fre-quency	Percen-tage	Fre-quency	Percen-tage	Fre-quency	Percent-age
TV	88	83.81	16	15.24	1	0.95
Radio	8	7.62	48	45.71	49	45.67
Newspaper	70	66.67	20	19.05	15	14.28
Magazines	7	6.66	49	46.67	49	46.67

Radio : It was seen that majority of radio users were not using radio regularly. Only 7.62 per cent of radio listeners used radio regularly followed by causal listeners (45.71%).

Newspaper : Out of the total newspaper readers 66.67% of the respondents were reading newspaper regularly. Only 19.05% of respondents were reading newspaper sometimes.

Magazines : Magazines were not regularly used by the respondents are only 6.66% of total magazines users were reading magazines regularly and majority of respondents (46.67%) were reading magazines sometimes.

These findings partly support the findings of the Srivastava et al. (1996) but are contradictory to the findings of Wakle et al. (1998).

Purpose of Utilization

Purpose of utilization differs with the demographic profile of the respondents. The purpose of using any media may be for news or getting information, entertainment, education or just for passing the leisure time. The purpose of using different mass-media was studied and the results have been presented in Table-7. In this case multiple response was allowed so that percentage may exceed to hundred.

Television : It is clear from the table-7 that majority of respondents (90.38%) used T.V. for entertainment purpose followed by news or information purpose (86.53%). Only 13.46% of respondents reported to view the television for education while the remaining 4.80% of respondents viewed TV for passing the leisure time.

Radio : Majority of respondents (94.64%) listened radio for entertainment followed by news or information (32.14%) and education (5.35%). No respondent reported to use the radio for passing the leisure time.

Newspaper : The data in the Table-7 shows that majority of respondents (98.88%) read newspaper for obtaining news or information followed by education (10%) and entertainment (4.44%). None of the readers reported to read newspaper for passing the leisure time.

Magazines : The purpose of reading magazines as indicated by the respondents shows that maximum number of respondents (60.71%) read magazines for the purpose of entertainment followed by education (48.21%), news or information 32.14% and passing the leisure time 17.85%).

A look on the review of literature indicates that results of the study partially support the findings of the Ingole et al. (1993) and Dharmadhikari et al. (1994).

Table-7 : Purpose of mass media utilization.

Mass media	News/ information		Entertainment		Education		Passing leisure time	
	F	%	F	%	F	%	F	%
TV N = 104	90	86.53	94	90.38	14	13.46	5	4.80
Radio N = 56	18	32.14	53	94.64	3	5.35	-	-
Newspaper N = 90	89	98.88	4	4.44	9	10.00	-	-
Magazines N = 56	18	32.14	34	60.71	27	48.21	10	17.85

Note :*Multiple response was allowed.

Time of Utilization of Mass Media (Segment of the day Utilized)

The segment of the day when people use media is important to know so that effective development programmes can be telecast or broadcast at appropriate time to seek maximum involvement of people. This will further help the development planners in meeting the development message needs of the people. Findings of the study have been presented in Table 8.

Television : Table-8 reveals that majority of the television viewers (55.76%) were watching T.V. at night (after 8 PM). However, 36.53% of the televiewers viewed T.V. in the afternoon (12-4 PM) followed by morning (33.65%) and evening (30.76%). Only 0.96% of the respondents were watching TV at time when they were free.

The findings indicate that night is the good time for disseminating information among farm families. At night, they are free from their daily chores.

Table-8 : Segment of the day spent on mass media utilization.

Mass media	Morning (6-12 noon)	Afternoon (12-4 PM)	Evening (4-8 PM)	Night after (8 PM)	Free time
TV N = 104	35 (33.65)	38 (36.53)	32 (30.76)	58 (55.76)	1 (0.96)
Radio N = 56	9 (16.07)	27 (48.21)	12 (21.42)	-	15 (26.78)
Newspaper N = 90	63 (70.00)	21 (17.77)	16 (17.77)	2 (2.22)	4 (4.44)
Magazines N = 56	5 (8.92)	10 (17.85)	11 (19.64)	5 (8.92)	37 (66.07)

Note : Figures in parenthesis are in percentage.

*Multiple response was allowed. So total of percentage may be more than 100.

Radio : Majority (48.21%) of respondents reported to listen radio in the afternoon followed by free time (26.78%), evening (21.42%) and morning hour (16.07%). Further it is evident from Table-8 that no respondent listened the radio at night. It may be due to fact that people in the villages used to watch T.V. at night for entertainment purposes. Therefore, they did not give attentionto radio programmes.

Newspaper : It is evident from the Table-8 that out of the total newspaper readers 70% read newspaper in the morning hours while a meager 2.22 % respondents read at night. However, 23.33%, 17.77% and 4.44% respondents reported to read newspaper in the afternoon, evening and free time respectively.

Magazines : The reading time of magazines as reported by the respondents indicated that most of readers (66.07%) read magazines whenever they were free. Thus, having no fixed time for magazine unlike other media the number of respondent reading magazine in the evening, afternoon, morning and night were to the tune of 19.64, 17.85, 8.92 and892% respectively. To some extent, these findings are similar to the findings reported by Natikar et al. (1995) but are dissimilar to the findings of Singh et al. (1999). **Duration of Mass Media Utilization.**

It is very essential to know the amount of time spent by the people with mass media so that we can find out which medium gets what share of time of an individual. Informationregarding this aspect will help the development planner to plan, prepare and tail or the development programmes according to an appropriate duration of time desired by the people. Data regarding this aspect was collected and have been presented in Table-9.

Table-9 : Duration of utilizing mass media.

Hours spend per day	TV N = 104	Radio N = 56	Newspaper N 90	Magazines N = 56
<1 hour/day	16 (15.39)	35 (62.50)	40 (44.44)	28 (50.00)
1-2 hours/day	35 (33.65)	18 (32.14)	46 (51.11)	7 (12.50)
>2 hours/day	53 (50.96)	3 (5.36)	4 (4.44)	21 (37.50)

Note : Figures in parenthesis are in percentage.

Television : It is evident from Table-9 that maximum number of television viewers (50.96%) devoted more than two hours per day while 33.65% viewers devoted time between one to two hours per day on T.V. viewing. However, 15.39% T.V. viewers reported to devote less than 1 hour in viewing television.

Radio : It is depressing to note that out of 105 respondents, only 56 were radio listeners. Further, it is revealed by the study that out of total radio listeners, majority (62.50%) devoted less than one hour per day on listening radio. Only 5.35% radio listeners reported to devote more than two hours per day on listening radio. It is a matter of further study to find out the factors behind this poor listenership and meager time devoted to this important media of communication.

Newspaper : It gives pleasure to note that out of 105 respondents 90 were newspaper readers. The present study further revealed that out of total newspaper readers, majority (51.11%) of readers devoted time between 1 to 2 hours per day. However , 4.44% readers reported to devote more than 2 hours per day on reading newspaper.

Magazines : The study indicates that magazine readers were less than newspaper readers. Fifty per cent of total magazine readers reported to devote less than one hour per day on reading magazines. Respondents devoted more than 2 hours per day and 1 to 2 hours per day were to the tune of 37.50% and 12.50% respectively.

On the whole, it can be said that majority of the respondents were T.V. users followed by newspaper users. People devoted more time on watching T.V. and reading newspaper as compared to radio listening and magazine reading. It also indicates that radio has lost its popularity among the farm families of selected area. However, newspaper has surprisingly got popularity among the farm families. Television being an audio-visual media of communication was at the top in the list of selected mass media in terms of popularity. These findings are dissimilar to the findings of the Srivastava et al. (1996).

Implications

The findings of the present study hold a great potential in the field of communication research as well as for planning and dissemination of development messages for rural families. Some of the implications of the present study have been given below.

- Development messages should be incorporated with entertainment programmes as it has been revealed by the study that people use T.V. and Radio mostly for the entertainment purpose.

- T.V. and newspaper should be used effectively for dissemination of development messages as it has been found that majority of the people use T.V. and newspaper regularly for larger period of time. It may be noted that even large percentage of non-media owners also use other's media for satisfying their information need.
- T.V. programme related to development messages should be telecast at night and radio programme should be broadcast in the afternoon as majority of the farm families indicated their favourite time.
- Need based and participatory approach should be adopted by media personnel and organization for programme production or message designing. Print media should use simple words and local terminology in publishing materials or information for rural farmers and farm families. Both electronic media and print media should use participatory message designing for development communication.
- Radio seems to loose attraction among the farmers of 'Tarai'. Radio station should try to make efforts like 'Radio FM' and 'Radio Mirchi' which are gaining popularity in the masses.

REFERENCES

1. Begum, Y. (2000). Designing print media messages on health and nutrition for rural mothers under ICDS. *Thesis, M.Sc.* G.B. Pant University of Agril. & Tech., Pantnagar.
2. Chauhan, J. and Singh, A.K. (2001). Use of communication sources by tribal dairy farmers- A study. Indian Research Journal of Extension Education, 1 (1): 8-10.
3. Dharmadhikari, S.D.; Sundaraswamy, B. and Shinde, P.S. (1994). Televiewing behaviour of T.V. owning farmers. Maharashatra Journal of Extension Education, Vol. 13 : 287-288.
4. Ingole, N.P. ; Saigaonkar, P.B. and Kotnekar, M.D. (1993). Television viewing behaviour of farmers. Indian Journal of Extension Education, 29 (3 & 4): 54-55.
5. Kukrety, N. (1990). A study on utilization of mass media and interpersonal channels for Agricultural Information in a hill village of Uttarkashi District of U.P. Thesis, M.Sc. G.B.P.U.A&T., Pantnagar.
6. Kumar, M. (2000). Communication of development messages among rural poor- An ethnographic study. Thesis, M.Sc., G.B. Pant University of Agril. & Tech., Pantnagar.

7. Nanjappa, D.; and Ganapalhy, K.R. (1996). Newspaper reading habit of farmers of Banglore district Karnataka. Indian Journal of Extension Education, 22 (1&2): 25.

8. Natikar, K.V.; Gavimath, U.S. and Budihal, R.A. (1995). Reading behaviour of the farmers of North Karnataka. Indian Journal of Extension Education, 31 (1 to 4): 87-90.

9. Pandey, R.K.. (1997). Information sources and communication pattern of farmers for rural development programmes : A study in U.P. Thesis, Ph.D., G.B.P.U.A.&T., Pantnagar.

10. Sangha, G.S. and Kalra R.K. (1993). Utilization of Mass Media sources by farmers. Indian Journal of Extension Education, 29 (1 & 2): 139-141.

11. Singh, D. (1997). A study of information sources of hill women of Charnba block in Tihari-Garhwal District of U.P. Thesis, M.Sc., G.B.P.U.A.&T., Pantnagar.

12. Singh, L.; Lal, R. and Chand, R. (1999). Radio listening behaviour and its impact on farmers. Agricultural Extension Review, 11 (4): 21-26.

13. Srivastava, A.; Dubey, V.K. and Mishra, S.K. (1996). Impact of mass-media on urban women of Varanasi, Interaction, 14 (2): 80-100.

14. Wakle, P.K.; Wattamwar, V.T. and Khalge, M.L. (1998). Utilization of different sources of farmers for seeking farm information. Maharashatra Journal of Extension Education, Vol. 17: 64-67.

Chapter **29**

Communication Pattern in a Slaughterhouse : A Case Study

Neeraj Kumar & P.N. Kaul

In spite of some social and political taboos associated with meat business Indian meat industry has got very good potential both within the country and abroad. According to an estimate, India has earned more than 375 crores of rupees by exporting meat in the year 1989-90 (FAO, 1990) and it is estimated that by end of the year 2000 the income from export will increase tenfold (Laxmanan, 1992). Slaughterhouses are the only authorized centres of meat production in the country. Nearly all the 36,000 registered slaughterhouses in the country are under government administration. Unfortunately, the status of these production centres of the country is abysmally miserable (Balain, 1993; Ramaswamy, 1994; Singh, 1994; Kapoor, 1994, Gandhi, 1994). Any attempt to ameliorate the status of slaughterhouses of the country will require basic information about the organization and their various components. However, the authors were unable to find out any effort to study the organizational component of such organizations which are of great economic value. In this backdrop, an attempt was made to study the communication pattern of the organization which is among the most important components of the organizational structure and function.

Communication is said to be at heart of any organization (Hall, 1987). In an exhaustive theory of organization, communication would occupy a central place, because the structure, extensiveness, and scope of the organization are almost entirely determined by communication techniques (Barnard, 1938). Probably the most salient and clearly evident aspect of structural organization within a group is the stabilization of particular lines of communication among individual members of the group (Me. David and Harari, 1986). Further, it has been found that different facets of communication viz., types of communication, direction of communication, content of communication, types of problems in communication and ways of solving them etc. get more or less stabilized with the time.

Organizations develop their own culture with language rituals and styles of communication (Kanter, 1977). Although, it being a dynamic process never attains static character but, it is true that by studying this dynamic process one could predict about the organization.

Research Methodology

To study the communication pattern of the slaughterhouse case study method was followed. A slaughterhouse somewhere in north India was purposively selected as the local of the study. Information was collected mainly through non-participant observation which was substantiated by informal interviews whenever required. Observation was made for total 40 days and it was assured that information is gathered from all possible corners of the slaughterhouse and all the communications were recorded in term of communication episodes and it was used as the unit of communication to analyze different element of the communication pattern. Communication pattern in the slaughterhouse was studied in terms of participation in communication, content of communication, type of communication, flow of communication and network of communication.

Results and Discussion Participation in Communication

This was studied in the term of level of participation by the different workers in the ongoing communication during the normal working hours of the slaughterhouse. Total 5,493 episodes were recorded in which all the participants were noticeable, The level of participation was worked out by finding out major participants in every communication episodes. In majority of cases, the communication was between two but,in some cases where more than two members were involved, two most active participants were considered.

As Table-1 indicates butchers were most active communicating group in the slaughterhouse followed by blood collectors. When considered two major participants in different communication episodes communicating each other it was observed that two butchers are involved in communication in maximum number of episodes. After butcher - butcher and butcher-blood collectors, next major communicating partners were retailers and middlemen. However, their communication was in the auction ground and not during the slaughtering process.

Such findings are probably because of the nature of work in which they were involved. During slaughtering butchers' work was main job and other work depended on it. Secondly, their nature of work demanded team-work viz., to get animal grounded, get it under full control before slaughtering, demanded support from additional two people and to get such support, the butcher had to communicate. Similar effect of nature of work on communication was also observed by Napier and Gershenfeld (1996) who reported that when people enter into a task with a predefined need to cooperative and interdependent, there is more listening, more acceptance of ideas, less possessiveness of ideas, and in general more communication. Third, being in the middle ofthe informal hierarchy and an important player in the entire process butchers had access to large group of people involved slaughterhouse-related activities. The effect of status on communication has also been explained by Shaw (1976). Fourth, the butcher was present both during the slaughtering period as well during the auctioning time, and hence, there was more possibility of communication. Even the number of butchers working in slaughterhouse was more than that of any group of workers. Further, those at the same level of hierarchy are more apt to share common characteristics making communication even more likely (Hall, 1987). Last, their physical spatial arrangement of work place was as such that there were other butchers working very close to them from either both sides or at least from one side and hence there was more communication. Explaining the effect of spatial arrangement Steinzor (1950) reported that members speak more to people seated directly opposite them than to those occupying adjacent seats. A similar result was reported by Stoodbeck and Hook (1961). Reasons for being most talkative pairs (butcher-butcher) followed by retailer middleman were related to the sites of their respective work and nature of work they were associated with.

Content of the Communication

To study the content of the communication, major contents of audible episodes were noted and were later classified into nine categories. Categories identified were: (a) ongoing job related (b) related to their past and future job (c) about colleague(s)[1] work (d) personal/family related (e) quality of animals/products (f) about market price of related products (g) about existing facilities/system (h) sports/movie/festivals. In total, therewere 811 episodes, in which the observer was able to delineate their major content.

Participants	Butcher	Blood Collector	Blood contractor	Blood churner	Cleaner	Auto waalas	Retailer	Gate keeper	Doctor	Middle man	Business man
Butcher	732	695	109	53	25	85	152	121	40	30	10
Blood collector	XX	305	142	207	12	95	50	167	17	07	XX
Blood contractor	XX	XX	XX	XX	XX	03	XX	10	XX	XX	XX
Cleaner	XX	XX	XX	XX	75	70	127	78	XX	XX	XX
Autowaalas	XX	XX	XX	XX	07	XX	25	61	32	240	35
Retailer	XX	XX	XX	XX	39	XX	XX	50	147	XX	95
Gatekeeper	XX	XX	XX	XX	XX	XX	XX	XX	XX	XX	70
Doctor	XX	XX	XX	XX	XX	XX	XX	XX	XX	54	XX
Middleman	XX	XX	XX	XX	XX	XX	XX	XX	XX	XX	75
Businessman	XX	XX	XX	XX	XX	XX	XX	XX	XX	XX	75

S. No.	Content of communication	Frequency
1.	Work (continued) related	331 (40.81)
2.	Work (Past/Future) related	157 (19.37)
3.	About colleague's work	055 (6.78)
4.	Personal/Family related	079 (9.75)
5.	About market price of products	032 (3.94)
6.	About existing facility and doctor	016 (1.97)
7.	About movies/sports/festivals etc.	094 (11.59)

Table-2 shows that people working in slaughterhouse were most concerned with their job and they talk most about their on going work (40.81%). The contents related to their future/past job were also talked about considerably. It is important to note that even during the work hours they talked about their non-job related activity e.g., sports, movies, festivals (11.59%). It was found that contents related to quality of animals came for discussion on three different occasions, viz., at the time of auctioning, at the time of animal verification, and during the slaughtering. But, when compared with other issues very less episodes were devoted to it. They seemed quite contented with existing system & facilities as they had least discussion about it.

As it is evident from the above discussion that people talked most about their on-going job. This is important to analyze because most of the workers resided in same locality and were related to each other by blood as well, and hence, there could have been sufficient common areas of experience which they could have discussed. Katz and Khan (1986) had concluded that if there were no problems of task coordination left to a group of peers the contents of their communication could take from which were irrelevant or destructive to organizational functioning. This may be because their work required continuous attention towards their work, and at work there was very less opportunity to think and discuss about non-work related issues. In other words, their job had a time limitation and hence they had very little time for leisurely discussion.

Types of Communication

Types of communication were studied in terms of mode of communication, i.e., verbal or non verbal, one to one communication or one to many communication. Verbal communication was further subdivided into oral communication and written communication. To study the non-verbal communication different clues like distance, posture, orientation, physical contact, facial expression, gestures, physical expression were identified. It is important to note that most of the oral communications were usually accompanied by non-verbal clues. However, for the present study non-verbal communication was considered only when at least one message was convened by only non-verbal mode. For analysis 811 episodes were considered. Table-3 indicates that about per cent of the communication was done through non-verbal mode of communication. Among the verbally communicated episodes the share of written mode was negligible. During the entire observation period only six written messages were related to the slaughtering activity and which has the bearing on butchers as well.

A significant number of communications (15.05%) were made from one to many individuals. Such one to many mode of communication was mostly followed in two situations (a) when butcher calls blood collectors before he slaughters the animal and (b) when middleman or his representative gives a call for bidding.

The forgoing discussion indicates that considerable number of communication was made through non-verbal mode of communication only. The process seemed well established and everybody understood it. Further, there as a continuous commotion at the working place and the job was manual and required continued interventions and attention; so, workers preferred non-verbal mode of communication to that of verbal. Similarly, one to many communication were because the messages were intended for many and the source never expected any particular individual to react to his message. Instead, it was open for all.

Table-3. Classification of communication episodes on the basis of types of communication.

a. Verbal and non-verbal.

b. Spoken and written communication

c. One to one and one to many communication

Table : a

Type of communication	Verbal	Non-verbal
Frequency, N = 881	617 (70.03)	264 (29.97)

Values in parentheses indicate percentage of total

Type : b

Type of communication	Spoken	Written
Frequency, N = 617	611 (99.02)	06 (0.08)

Values in parentheses indicate percentage of total

Table : C

Type of communication	One to one	One to many
Frequency, N = 881	689 (84.95)	122 (15.05)

Values in parentheses indicate percentage of total

Communication Networks

Communication networks were studied in terms of regular pattern of person to person contact that are characterized by the exchange of information between them.

1. Based on the observation a regular pattern of communication was delineated. On a normal working day in the slaughterhouse during the working hours communication was mostly limited to that between or among those who worked together as an unit. The unit usually comprised of a butcher head and associate butcher(s) (one or two). However, blood collectors were regular intruders in their network and they also participated till they stayed on the site for either helping them or for blood collection. Butchers also communicated to their next neighbours mostly when they needed some help or they wanted to relax. In the end, auto-waalas and retailers joined the network, although for shorter duration but communication was very intensive.
2. Another network comprised to gatekeeper, blood collectors, contractor, auto-waala and retailers. While keeping watch on the animals' movement, the gatekeeper usually entered in communication with those taking animals out. Blood contractor while standing near the gatekeeper and keeping watch on blood collection, frequently communicated with the gatekeeper. During the end of working hours, auto-waalas and retailer who came for collecting meat got involved in small conversation with gatekeeper.
1. Third obvious network in the slaughterhouse semed to be in blood processing room. Churners {most of the time 3 in number) were always found involved in interpersonal communication and blood collectors who came with blood often interrupted

the communication with their questions or other comments. As soon as blood collectors left the room they continued with the original issue. Another participants in network was the contractor who after entering the room influenced the ongoing communication.

2. Just before auctioning started retailers (potential buyers} frequently interacted with fellow retailers or middlemen. Middlemen who reached the auction site a little late also interacted with retailer, fellow middlemen and also with the businessmen- It was observed that all the businessmen except on three occasions got their animals sold through respective middlemen. Hence, middlemen had communication restricted to those businessmen who were their clients.

The communication network was open for all and everyone could directly contact others, if desired. However, because of nature, time and place of different work some regular pattern of communication was observed resulting in networks. The first network comprising butchers, blood collectors was formed during the slaughtering and butcher head was in the centre of this network. This network was formed because of specific work and involvement of concerned individuals. The effect of specific work has also been explained by Mitchell (1978) who concluded that increasing specialization of the work force hindered communication flow as when every one had a special job it became increasingly difficult to see and talk to people inside one's speciality. In another network with gatekeeper in the centre, blood contractor, auto-waalas and retailers were members. This network was because of the specific place where gatekeepers used to stand. Festinger et. al. (1950) and Shaw (1976) too had noted effect of housing units of friendship formation and influence of spatial arrangements on group differentiation respectively which have obvious effect on communication. The third network with churners in central position was also because of their specific work and place, as churners were working in a particular room where only those related with blood processing work went and formed the network. The last network was work and time specific.

Table-4. Classification of communication episodes according to their direction

Direction	Downward	Horizontal	Upward
Frequency, N = 881	441 (42.00)	336 (32.00)	273 (26.00)

Values in parentheses indicate percentage of total

Flow of Communication

Flow of communication refers to the direction of flow of communication. In an organization flow of communication may be from management or higher authority to his subordinate, among the workers of same level or from subordinate to higher authority. These are known as downward communication, horizontal communication and upward communication respectively.

To study the flow of communication, communication episodes were identified as units. In each communication episodes the direction of flow of communication was noted.

In total 1050 episodes were recorded where the observer was able to find out the direction. It was found that 42 percent of episodes started from the seniors and were directed towards juniors. 32 percent of communication had horizontal and the rest 26% had upward direction of communication.

Downward Communication

As Table-5 speaks majority of communication started from the main butchers and was directed towards associate butchers and blood collectors (35.10% and 14.10% respectively). Communication between blood contractor and blood collectors (15.80%) and between gatekeeper and butcher/assoc. Butcher (11.10%) took place during the slaughtering process only. Doctor had communication with gatekeeper during the day time or during the animal verification. During the day time (usually twice a week), doctor used to take stock of day-to-day functioning and then instruct him accordingly. He discussed the butchers' complaints if any, which he had received during the animal verification on previous day(s). Retailer to butcher communication which constituted 10.3 per cent to total downward communication was made during two occasions viz., during the auction and/or at the end of the slaughtering process.

Table-5. Different participants and their respective share in downward communication

S. No.	Parties involved	No. of episodes
a.	Deptt. to all concerned	05 (1.10)
b.	Butcher heads to Assooc. Butchers	155 (35.10)
c.	Butcher to blood collectors	602 (14.10)
d.	Blood contractor to Bl. Collector	70 (15.90)
e.	Gatekeepers to butchers	49 (11.10)
f.	Doctor to gatekeepers	55 (12.50)
g.	Retailers to butchers	45 (10.30)

Values in parentheses indicate percentage of total

The messages regarding the official closure of the slaughterhouse due to festival (04) and repair of roof-top (01) had come through official letter. Out of these five such letters which had information/ order regarding the closure of slaughter -house two were shown to gatekeeper and subsequently he was asked to convey this to all concerned workers. On two occasions gatekeepers had brought notices from the district office, and once doctor verbally told gatekeeper about the closure of slaughterhouse mentioning that he had received the letter. It is important to note that on most of the occasions doctor had interacted with one particular gatekeeper.

Table-6. Different participants and their respective share in horizontal communication

S. No.	Parties involved	No. of episodes
a.	Butcher to butcher	94 (28.00)
b.	Retailer to middlemen/retailer	62 (18.50)
c.	Blood churner to Bl. Chuner	120 (35.70)
d.	Gatekeeper to butchers	25 (7.40)
e.	Gatekeeper to blood contractor	25 (7.40)

Values in parentheses indicate percentage of total

Horizontal Communication

Communication at same level took place between butcher and butcher, butcher and gatekeeper, churner and churner, gatekeeper and blood contractor, retailer and retailer, and between retailer and middleman (Table-6). It is clear that maximum horizontal communication took place among blood churners themselves (35.70%)- Blood churners who remained confined inside a room had continuous talk even when they were on work. Further, there were minimum disturbances in their work. Butcher to butcher communication (28.00%) which started just after their arrival continued till they finally left the slaughterhouse. It is important to note that communication between butchers and associate butchers has also been discussed under the downward communication. It is because associate butcher although, worked under main butcher but, back home they live in same locality and were related to each other by blood. Another important finding is that butcher while communicating something related to ongoing work communicated with authority and when message is not related to work it was like friendly talks. Horizontal communication between retailers and middlemen (18.5%) took place before and during auctioning process. Communication involving gatekeepers i.e., gatekeepers - blood contractor (10.40%) and gate keeper - butchers (97.40%) was mainly because of position of gatekeeper where he stood. It was convenient for blood contractor to supervise the blood collection work by standing near to him and, hence, communication took place. Similarly, as butchers had to go to stable to bring the animals out or to go out for some work as he was on the way the communication between two materialized.

Table-7. Different participants and their respective share in upward communication

S. No.	Parties involved	No. of episodes
a.	Gatekeeper to doctor	67 (24.54)
b.	Blood collector to contactor	36 (13.19)
c.	Blood collector to butcher	76 (27.84)
d.	Butcher/retailer to doctor	42 (15.38)
e.	Cleaner to gatekeeper	52 (19.05)

Values in parentheses indicate percentage of total

Upward Communication

Blood collection being the most mobile and directly related to the work of slaughtering made maximum upward communication (41.00%). As blood collectors had the responsibility of extending helping hands to butchers in slaughtering process they had considerable communication (27.80%) with butchers (Table-7). Blood contractors, to whom butchers were responsible, received more than thirteen percent of upward communication. Veterinary doctor, who is overall incharge of slaughterhouse, was addressed respectfully during maximum number of episodes (about 40 percent). However, these were gatekeepers who had most (61.46%) of such communication addressed to doctor. Butcher or retailers had communication during the animals' verification. Butchers or retailers have been considered together as they had the same purpose of communication and there was not fixed pattern of their participation during the animals' verification. For cleaners a gatekeepers was like their reporting officer, and hence, the communication was like a subordinate to superior. It is important to mention that most of the communication involving gatekeepers was done by one particular gatekeeper which were reality five major reasons of downward communication. It was also observed that many of the downward communication were neither made as 'orders' nor they were taken so by subordinates. This supports the thesis put forward by Hall (1987) who concluded that when hierarchy is composed of people of equal or higher ability than the subordinates their function in this regard is clear. They can make suggestions and offer alternatives that might not be apparent to subordinates.

The communication but the peer group is a common feature of horizontal communication but the communication between butchers and associate butchers was observed both under the downward communication and horizontal communication. This indicates that flow of communication also depended upon the context in which communication took place irrespective of status of participating partners. Wager (1972) and Robert et al. (1978) had found that the role that the individual plays in the organization affects how communications are perceived or sent. After reviewing many studies on this aspect, Hall (1978 had concluded that in almost all organizations people can be superordinate in one situation and subordinate in others. Considerable number of communication having upward flow can be attributed to the nature of hierarchy where superordinate had very little power to affect their respective subordinate and hence, they felt free in communicating upwards.

Conclusions

Based on the above description the following conclusions can be drawn : Participation in communication is dependent on the nature of work, one's social status in the existing social hierarchy, spatial arrangement at one's work place, participation in activities involving different groups of people, and number of peer group members present during the possible communication situation. If the work requires team-work and individuals can not accomplish their respective works on their own involvement of workers in communication is more likely. Being in the centre of social hierarchy provides comparatively access and opportunity of communication with people from both sides of status hierarchy. If the spatial arrangement of working place of an worker is as such that it provides physical greater access to many other people the possibility of entering in communication of the worker is quite high. The presence of more number of peer group members and presence of one group of people during the different activities naturally increases their share in overall communication. Content of communication depends upon the nature of work. If the work demands continuous involvement in job, workers could talk about their ongoing job only. Non-verbal communication not only complements or gives strength to verbal communication but, in certain circumstances it can replace it as well. When situation is not conducive (e.g. a lot of commotion and participants are at considerable distance) people opt for non-verbal mode of communication. Formation of communication network is contingent upon nature of work, place of work, and time of the work. The direction of flow of communication is dependent upon characteristics of organization like existence of explicit social hierarchy, nature and procedure of work, and context of communication. It is important to note that flow of communication depended upon context of communication irrespective of status of parties involved.

Although above conclusions have been drawn from the findings of study of one organization and subsequent discussion corroborating those findings, these (conclusions) do bring forth some plausible hypothesis. These hypothesis are important because of different nature of the organization studied and need to be tested in varied conditions for possible generalization.

REFERENCES

1. Barnard, C.J. (1938) The Functions of the Executive, Cambridge, M.A.; Harvard University Press.
2. Balain, D.S.; Lakshmanan, V.; Anjaneyulu, A.S.R. and Yadav.
3. P.K. (1993) Buffalo for Meat: A Panorama, An unpublished paper, Indian Veterinary Research Institute, Izatnagar.
4. FAO (1990) Production year book, FAO, Rome.
5. Festinger, L. (1950) Informal Social Communication, Psychological Review, 57, 271-282.
6. Gandhi, Maneka (1994) Criminal Record of Idgah Abattoir, Times of India, dt 25th April'94, New Delhi.
7. Hall, R.H. (1987) Organizations Structures, Process and Outcome, 4th1 Edition, New Jersy, Prentic Hall Inc.
8. Kanter, R.H. (1977) Men and Women of the corporation, New York, Basic Books.
9. Kapoor, Aditi (1994) Waste disposal still neglected at Idgah. Times of India, dt. 6th April'94, New Delhi.
10. Katz, D. and Khan, R.L. (1966) The Social Psychology of Organization, New York, John Wiley and Sons.
11. Kelly, H.H. (1951) Communication in experimentally Created hierarchies, Human Relations, 4, 39-56.

12. Lakshmanan, V. (1992) Present status of Salvaging male buffalo calves for meat production: A field study, Unpublished Ph. D. Thesis, Indian Veterinary Research Institute, Izatnagar.

13. Luthans, F and Larsen, J.K. (1986) How managers really communicate, Human Relations, 39, 161-178.

14. McDavid, John W. And Harari, Herbert (1986) Group structure and the Individual, Social Psychology, Individual, Groups Soceities, New Delhi CBS, Publication.

15. Mitchell, Terence R. (1978) Communication, People in Organization Understanding their Behaviour, New Delhi, McGraw Hill Book Company.

16. Napier, R. W. And Gershenfeld, M.K. (1996) Perception and communication, Groups: Theory and Experience, 5'" Edition, New Delhi, AITBs.

17. Ramaswamy N.S. (1994) Shocking Conditions in Abattoirs, Times of India, dt. 4th April 94, New Delhi.

18. Roberts, K.H.; Hulin, C. and Rousseau, D.M. (1978) Developing an Interdisciplinary Science of Organizations, San Francisco, Jossey-Bass.

19. Shaw, M.E. (1976) Group Dynamics: The Psychology of Group Behaviour, 3rd Edition, New Delhi, Me. Graw Hill Book Company.

20. Singh, N. (1994) Hamaare Jivan Rakshak Hain, Pashu, Nav Bharat Times, New Delhi.

21. Steinzor, B. (1950) The spatial factor in face discussion groups, Journal of Abnormal and Social Psychology, 45, 552-555.

22. Stoodbeck, F.L. and Hook, L. (1961) The social dimensions of a twelve man jury table, Sociometry, 24, 397-278.

23. Thibaut, J.W. (1950) An experimental study of the cohesiveness of under privileged groups, Human Relations, 3, 251-278.

24. Wager, L. Wesley (1972) Organizational 'Linking - pins': Hierarchical status and communicative poles in Inter-level conferences, Human Relations, 4, 307-26.

25. Walker, C.R. and Guest, R. H. (1952) The man on Assembly line, Harvard University Press, Cambridge, Mass.

Chapter **30**

Agricultural Knowledge Information System- An Emerging Communication Approach for Sustainable Development

Anindita Saha, Jancy Gupta, Swapan Behari Mukhopadhyai

A good number of communication approaches are available for agricultural development. Present orientations, such as Farming System Research (FSR) and Training and Visit (T&V), extension are unable to achieve synergy as they are centralized, default to wealthier rural elite, do not build farmer's capacity and ignore informal demands. Looking at farmers only as users neglects the important fact that farmers are experimenters themselves and that farmers have developed most of the technology used on farm today. Farmers do adopt ready-made technology, but they are also active problem solvers who create and share solutions to those problems in horizontal network. A serious technology mobilizing efforts can not afford to neglect the collective creative capacity of farmers (Rolling and Engel, 1991). Hence, system management in the knowledge system will be required to hold all formal and informal elements together, coordinate interactions, build concerns and monitor outcomes of synergy (Walker, 1994). Earlier farming system, extension, agricultural technology development research and policy making were considered as separate entities each with its own set of issues, managed by groups of researchers and professionals. There is a conceptual progression from looking at various institutions and practices such as farming system development, extension and research in isolation to considering the linkages between pairs of these elements and now to looking at all the elements as an Agricultural Knowledge Information System (AKIS).

Genesis of Agricultural Knowldge Information System (AKIS)

Generation of innovation is the responsibility of both the researchers and extensionist : the former should engage in 'research for future', the later in 'research of immediate applicability'. Such a distinction is not only untenable, but must lead to duplication of efforts and resultant inefficiency. If the work of research and extension is to make the maximum contribution to sustainability needs a shift from 'imposed technologies' to 'demand driven' one (Ramkumar, S. 1995).

The concept of 'Agricultural Information System' was first introduced and explained by "UNe Jens Nagel" of the Technical University of Berlin in 1980. An Agricultural knowledge Information system is a system in which agricultural information is generated, transformed, transferred, consolidated, received, and utilized and feedback in such a manner that these process function synergically to underpin knowledge utilization by agricultural producers (Roling, 1989). Effective technology development and transfer depends on an interactive holistic system that Roling (1989) called the 'Agricultural Information System' which includes: a research sub system; a dissemination sub system; auser subsystem. Roling and Engel (1991) define AKIS as "a set of organizations and/or persons, and the links and interaction between them that are engaged in, or managed such processes as the anticipation, generation, transformation, transmission, storage, retrieval, integration, diffusion, utilization of agricultural knowledge and information, which potentially work synergistically to support decision making, problem solving and innovation in agriculture on a domain thereof.

Analytical Models of AKIS

Several analytical models have been developed to examine and assess AKIS. Swanson (1985) described the analytical framework developed by the International programme for Agricultural Knowledge System (1NTERPAKS) to examine and assess Agricultural Knowedge Systems.

The four major components of the analytical framework is defined by the INTERPAKS framework were :

- Policy, which included the external factors that directly, affected the technology system, including the utilization of technology by farmers.

  Technology development, which included the part of the

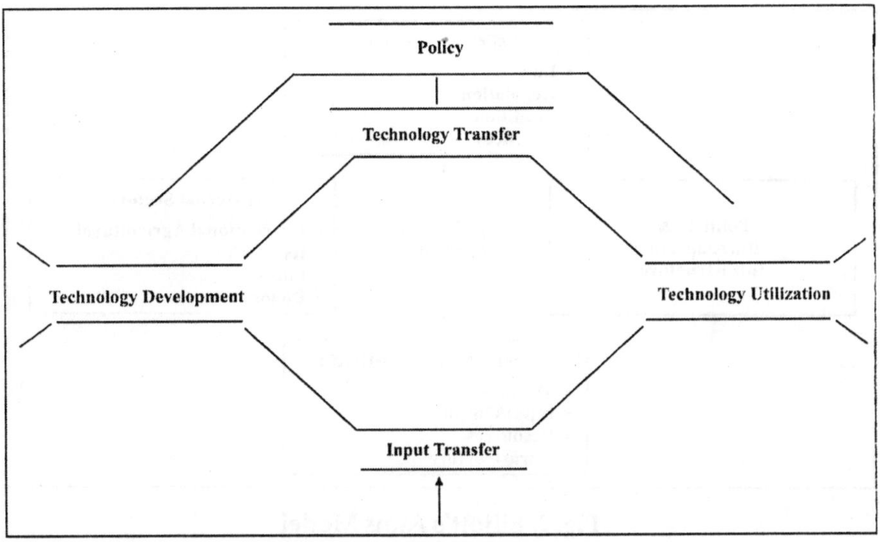

Fig. The Interpaks Model (Swanson, 1985)

agricultural research system that, was devoted to applied and adaptive research-

- Technology transfer, which was broken down into sub-function of knowledge transfer and input transfer.
- Technology utilization of farmers, with emphasis on small holders.

The synergic functioning of extension, research and other components of the AKIS were taken care by INTERPAKS. The INTERPAKS model depicted he AKIS as a one-way linear system (Roling, 1990). However, according to Kaimowitz et al. (1990) innovations which were policy driven were equally, if not more, important than science based technical innovation.

Elliott (1987) has proposed a model of the agricultural technology management system (ATMS) which takes the knowledge, information and technology factors and if there influence on the Agricultural Knowledge Information System into accounts in his model, he refers the technology sector what we have called AKIS. The technology sector contains the technology utilizing system, the technology generating system, and the technology transfer system-Roling (1989) has replaced Elliot technology sector with Agricultural Knowledge Information System.

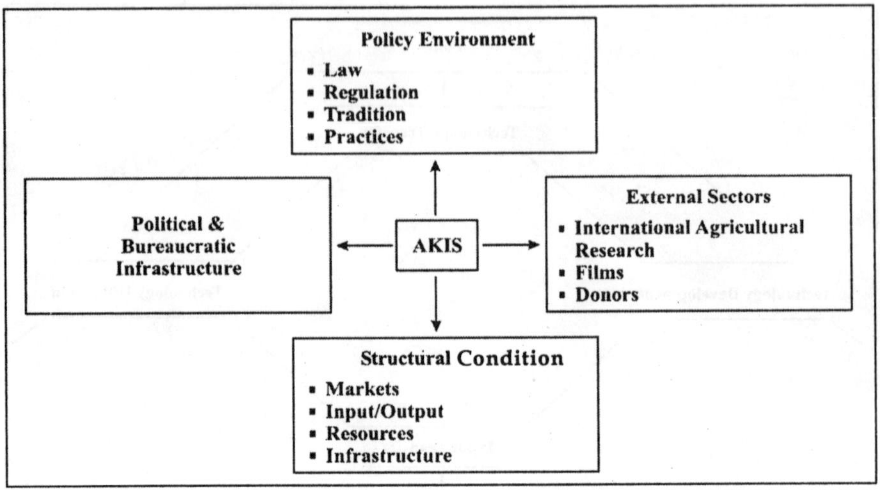

Fig. 2. Elliott's Atms Model

The Agricultural Technology Management System models of the AKIS, which did not reflect same flow of information and influence from technology users to other parts of the system, were misleading. This was not because of the moral principle that participation was needed if the system was to function benignly. It was because the empirical evidence from effective AKIS clearly demonstrated that user control in some form was an essential ingredient, participants in. effective AKIS had do take the needs of users into account at each point at which information or technology is transformed or adopted (Roling, 1989).

Elements of AKIS:

The main elements of this system are -

- Agricultural research that develops methods and new practices, as well as their modification in compliance with local conditions.
- Agricultural education that trains the extension specialist and directly assist the short-term training of the farmers.
- Organization of extension services of the central, regional and local levels, all separated from the regulatory function of the Ministry of Agriculture.
- The rural farming community.
- Mass media, through which information is spread, may be considered as a sub-element of this system. (Lefter Daku, 1997).

Roling and Engel (1991) gave a conceptual framework for AKIS based on a system approach. It was realized that such actors as research technology development, extension and farmers formed a whole, with emergent properties over and above those of a mere collection of actors. The combined contribution of these actors is more than the sum of the individual contributors. Policy, research, technology development, extension and farmers should be seen as a system.

Analysis of AKIS

It is important to analyze the AKIS for a certain region or a certain branch of agricultural in order to discover possible gaps, which hamper agricultural development, as well as possible overlaps, which waste resources and might cause conflicts. For example, policy makers are unable to design effective policies and extension programme for lack of adequate information about the actual situation on farms or performing the same task by private and government extension organization by spending money. This type of AKIS analysis can be the responsibility of the extension service. It should be used to improve the coordination between the actors in the AKIS by discussing how each organization can contribute as much as possible to make the AKIS more effective. Farmer's organisation also can play a vital role in the process by formulating the information needs of their members and stimulating research institutes, extension services and other actors to provide this information.

Engel (1995) developed a methodology to involve the actors themselves in the analysis of their AKIS : the Rapid Appraisal of Agricultural knowledge in the systems. This can be the starting point for a cooperative process to make this AKIS more effective.

It to especially important to conduct such an analysis of the AKIS before starting a new project.

When analysing the AKIS attention should be given to the fact that different groups of farmers such as male and female farmers, crop and livestock fanners and farmers with different levels of resources, possess and require different kinds of information and use different information sources.

The Role of AKIS in Sustainable Development

The traditional role of the extension organisation in less industrialized countries is the transfer of technologies (TOT) developed at the research institutes to farmers. A major role in industrialized countries has always been to learn form the experience of the most successful farmers in order to teach other farmers how they can improve their farm management. Quite often analysis of the AKIS shows that a role other than TOT is more appropriate. This could involve providing farmers with a basket of opportunities and helping them to choose the right one for their situation. Another role of the extension organization can be :

- To help farmers' experiment with new technologies or with new farming system;
- To gain access to relevant information from a variety of information.
- To evaluate and interpret this information for their own situation; and
- To learn from their experiences.

Development of a network to exchange information among all relevant actors is an important aspect here. Mc. Dermott (1987) has shown that it is usually necessary to integrate the information from researchers, farmers and extension agents to be able to develop technologies that work well in a given situation.

Nagel (1980) deliberately designed six functions that the system must perform in order to ensure the initiation and the continuation of the information flow process;

- Identification of problems/knowledge needs at the producers level;
- Generation of innovations;
- Validation under farmers' conditions/operanonalisation for utilization;
- Dissemination;
- Utilization;
- Evaluation of experiences.

AKIS framework is useful in analyzing how institutions involved in the process of generating and diffusion of new knowledge, i.e. research, extension and education, support farmers. The underlying feature of the AKIS is that farmers obtain the knowledge and information from many different sources and that new knowledge is generated not only by research institutes, but also by many different actors, including farmers themselves. Agricultural development will be more successful if the different actors in the AKIS share their knowledge and stimulate each other to develop new ideas about the direction in which agriculture should develop, and on the way this development can be achieved.

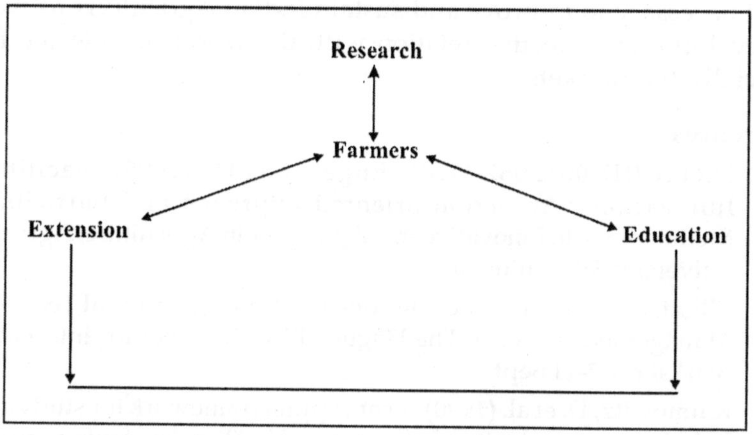

Sustainable development is the development that dosen't lost. Building an integrated system of research, extension, education and farmers community is considered to be one of the key ingredients for promoting agricultural development. Related to this, strengthening the linkages among element of AKIS constitutes a necessary condition for the well functioning of the system. Within AKIS, farmers not only are the central element, but also contributed to the improvement of research and extension work, and to the policy formulation process.

Conclusion

The development experience in the past decades, in spite of its progressive modification in the approaches and deliberations, has been still disappointing and distressing. This is glaringly true in the context of agricultural development in India. It has been experienced by all concerned with the agricultural development process that a significant gap exists between technology generation and technology use at the field level. The dual concept of AKIS and synergy can fill this analytical gap. Within AKIS, farmers are in the central position. Farmer's knowledge and skills are certainly an important complement to formal research and extension and such complementarily can be achieved through the institutional mechanism of AKIS. Farmers' organizations also can play a valuable role in the process by formulating the information needs of their members and stimulating research, institutes, extension services and other actors to provide this information. Strong intervention power and a good calibration of the research/practice continum do not only characterize successful AKIS, but also by user control. So, the articulated approach of AKIS is felt necessary to improve and sustain Indian Agriculture as well as to develop more effective relation with the world outside the farm, especially the markets.

References

1. DR11CRD-0012082-DGW Engel, P.G.H. (1995). Facilitating Innovation : An action oriented approach and participatory Methodology to Innovative social practices in Agriculture, Agriculture university, Wageningen.
2. Elliott, H. (1987). Diagnosing constraints in Agricultural Technology Management systems. The Hague : ISNAR, paper for international workshop, 7-11 Sept.
3. Kaimowitz, D. et al. (1990) A conceptual framework for studying the links between agricultural research and TOT in developing countries. West view press bulletin.
4. Lefter Daku. (1997). Investing in Agricultural Extension : The case of Albania.
5. Nagel, V. Z. (1980). Institutionalization of knowledge flows : An analysis of the extension role of two agricultural universities in India. Quarterly published Journal of International Agriculture, special volume series no. 30.
6. Ramkumar, S. (1995). 'The analysis of fanner information systems for feeding of dairy cattle in two villages of Kerala state, India'. Ph. D. Thesis. U.K. : The University of Reading.
7. Roling, N. (1989). The KIS : Content for IT. an attempt to draw the implications for information technology in extension from recent conceptualization about agricultural knowledge and information system, proceedings of a workshop held in London.
8. Roling, N. (1990). The Research - Technology Transfer Interface : A knowledge system perspective.
9. Roling, N and Engel, P.G.H. (1991). The development of the concept of Agricultural Knowledge Information Systems (AKIS) : Implication for Extension.
10. Swanson, B.E. (1985). 'INTERPAKS' and the development of its knowledge system model. International Agricultural Centre, Wageningen.
11. Walker, M. (1994). Agricultural Information Systems : Synergy and development. Journal of Extension System, 10(2).

Chapter **31**

Q-Sort Technique in Communication Research

A. Ramesh Babu and Y. P. Singh

The scaling procedure has become known as the 'Q-sort' technique, because of the feature of item sorting. The pre-fixing letter 'Q' has no special significance; by historical accident, the method came to be identified this way. The method was originally developed by Stephenson (1936) to provide, in convenient form, data suitable for his heuristic studies. According to catell (1944), the Q-sort method is an ipsative procedure i.e., the personality variables in the defined set are ordered or scaled relative to each other, with respect to a specified criterion and with a specific subject as the frame of reference.

In late 1952, a memorandum was prepared which presented the case for the Q-sort procedure as a means of codifying personality formulations in an assessment setting (Block, 1952). This memorandum led to the preparation of a Q-set for use by investigators (Mac-Kinnon *et.al.* 1958 and Block, 1961). Later the Q-sort technique was used by several investigators some of those were Marsh (1963). Kerlinger (1964). Nunnally (1967) and sontag (1963). With in the Extension research tradition in India, it has been used by Jaiswal (1965). Jha (1968), Girdar (1968), Jalihal (1970) and Ramesh Babu and Singh (1984). The Q-sort technique is described below :

I. **Item collection** : Items relating to the research topic, e.g. "Variables relating to Communication Effectiveness" can be collected through the following sources : (a) Review of earlier studies and (b) Interviews with experts. As many as 200 variables can be identified through the sources mentioned above.

II. **Item selection** : The 200 variables can be critically examined with a view to detect ambiguity, repetitiveness and rephrase the items wherever necessary. Let us assume 80 variables are selected finally for the use of Q-sort technique.

III. Use of Q-sort technique : The detailed procedure used in developing the Q-sort technique included the following steps :

- Deciding the number of items to be included,
- Use of cards.
- Choosing the type of sorting.
- Fixing up the number of categories to be used.
- Deciding the shape of distribution.
- Obtaining the reliability of the Q-sort.
- Getting the Q-sorts from a sample of judges.
- Scoring the data and
- Interpretation

The methodology underlying the above mentioned steps is briefly outlined below.

1. **Deciding the number of items to be included :** The number of items in a Q-sort technique is determined by convenience and statistical demands. For statistical stability and reliability, the number should not be less than 60 (40 or 50 in some rare cases). A good range is from 60 to 90 items (Kerlinger, 1964).
2. **Use of cards :** Conventionally, the Q-items are printed separately on cards, which permits easy arrangement and re-arrangement of the items until the desired ordering is obtained. The 80 selected items can be printed neatly on colour cards of size 3½" x 2", in order to facilitate the sorting by the respondents.
3. **Choosing the type of sorting :** There are two types of sorting : (a) unforced Q-sorting and (b) Forced Q-sorting.

The respondents in 'unforced Q-sorting' are free to place any number of cards in a particular category. On the other hand, the subjects in 'Forced Q-sorting' are instructed to place a specified number of cards in each category. According to Livson and Nichols (1956), Block (1961) and Kerlinger (1964) the forced Q-sorting has definite advantages.

Block (1961) on the relative merits of the 'freely sorted' versus 'forced-sorted' distribution of Q-items made the following observations :

1. The unforced Q-sorting procedure obscures recognition of the correspondences existing among evaluations whereas the forced

Q-sorting procedure permits a clear assessment of degree of equivalence.

2. The unforced Q-sorting tends to provide fewer discriminations than the forced Q-sorting procedure.
3. The unforced Q-sorting is not more reliable than is the forced Q-sorting procedure, even though with the latter procedure Judges are required to make discriminations they otherwise are inclined not to offer.
4. The unforced Q-sorting procedure provides data which is unwieldy and at times impossible to work with, whereas the forced Q-sorting procedure provides data in a convenient and readily processed form.

It would be ideal to use 'forced Q-sorting procedure keeping in view its definite advantage over the unforced Q-sorting' procedure.

4. Fixing up the number of categories to be used : The number of categories to be employed along with the judgmental continuum has to be decided. Obviously, more number of categories is better, since it provides more discrimination. But use of too many categories might pressurise the judge to the point where he responds with great difficulty and great randomness. Block (1961) stated that :

"In our own experience, and after some trial and error we have settled upon nine categories as sufficient and yet not too much".

Stephenson (1953) had used as many as 13. Nevertheless, Block emphasised that even though a nine interval continuum, makes a heavier demand on the judges, his experience suggested that, it elicits reliable discriminations. In view of the above, it would be better to fix upon nine categories for the Q-distribution.

5. Deciding the shape of distribution : The next decision had to do with the essential shape of the Q-distribution. According to Block (1961), the limiting possibilities for a symmetrical Q-distribution were;

(a) Unimodel distribution

(b) Rectangular distribution, and

(c) U-shaped distribution.

In a unimodal distribution, there is piling up of items in the middle categories of the continuum; in a rectangular distribution, there are an equal number of items in each of the categories; and in a U-shaped distribution, there is a predominance of items in the extreme categories with a de-emphasis of item placement in the middle categories.

Although, a rectangular Q-distribution provides more discriminations. Block (1961) has listed four reasons why a Q-distribution should deviate some what away from rectangularity towards a unimodel distribution These are :

(a) The one study on this issue (Livson and Nichols, 1956) had shown a somewhat unimodel distribution to be preferred, on the average, by a set of judges.

(b) Items placed in middle categories are less important than extremely placed items.

(c) Items placed in the middle categories represent most difficult and time consuming judgments for the sorter to make.

(d) The conventional indices used toexpress the similarity between Q-sorts pay rather little attention to discriminations made in the mid ranges.

It would be ideal to use a unimodel distribution recognising the above indicated merits. The shape of the distribution is shown in Fig. 1. The number of items to be placed in each category is as follows :

S. No.	Category	Number of items to be placed
1.	Most important	4
2.	Highly important	6
3.	Very important	9
4.	Quite important	13
5.	Somewhat important	16
6.	Slightly important	13
7.	Of little important	9
8.	Of very little importance	6
9.	Least important	7
	Total	**80**

For instructions given to a respondent see Ramesh Babu and Singh (1984). These instructions provide a brief orientation to the procedure, its purpose and suggest a convenient way of carrying out the task.

6. **Obtaining the reliability of the Q-sort :** The reliability of the data obtained from Q-sort technique can be worked out by the 'test-retest method'. The selected items can be subjected to Q-sorting twice from 10 competent judges, who are different from

the main sample of the study, at an interval of 15 days. The calculated correlation coefficient between these two sets of scores will indicate the level of reliability of the Q-sort.

7. **Getting the Q-sorts from a sample of judges :** pre-testing of the Q-sort technique can be done with sample of five judges. A set of cards containing the selected items and code numbers, instruction sheet and rank sheet will be given to each of the five judges for Q-sorting. They can be also personally interviewed by the investigator, to ascertain their views, before the actual printing of cards, instruction sheet and rank sheet.

8. **Scoring of the data obtained from Q-sort technique :** There are nine categories in which the respondents will be asked to sort out the 80 items, based on their importance. The categories can be assigned the scores ranging from 0-8. The score and range of each individual category are given as below :

Category	Score	Range
Most important	8	7.5 to 8.5
Highly important	7	6.5 to 7.5
Very important	6	5.5 to 6.5
Quite important	5	4.5 to 5.5
Somewhat important	4	3.5 to 4.5
Slightly important	3	2.5 to 3.5
Of little importance	2	1.5 to 2.5
Of very little importance	1	0.5 to 1.5
Least important	0	0.5 to 0.5

9. **Interpretation :** Items which received a median score 4.5 i.e., Quite important category and above will be considered as relatively important. The remaining items whose median score is falling below 4.5 will be considered as relatively less important.

One useful improvement affected in operating Q-sort technique through mail is the design of a printed 'Pusa Rank Sheat' to be mailed along with the cards and instruction sheet. The respondents will be asked to enter the code numbers of the items on the blocks provided in the rank sheet. Instead of cards, the subjects will be asked to mail back only the rank sheet in the enclosed envelope. This will reduce 40 per cent of the mailing expenditure and save a lot of time and labour during the analysis of data. It also greatly minimises the mailing inconvenience of respondents. For further information, see Ramesh Babu and Singh (1984).

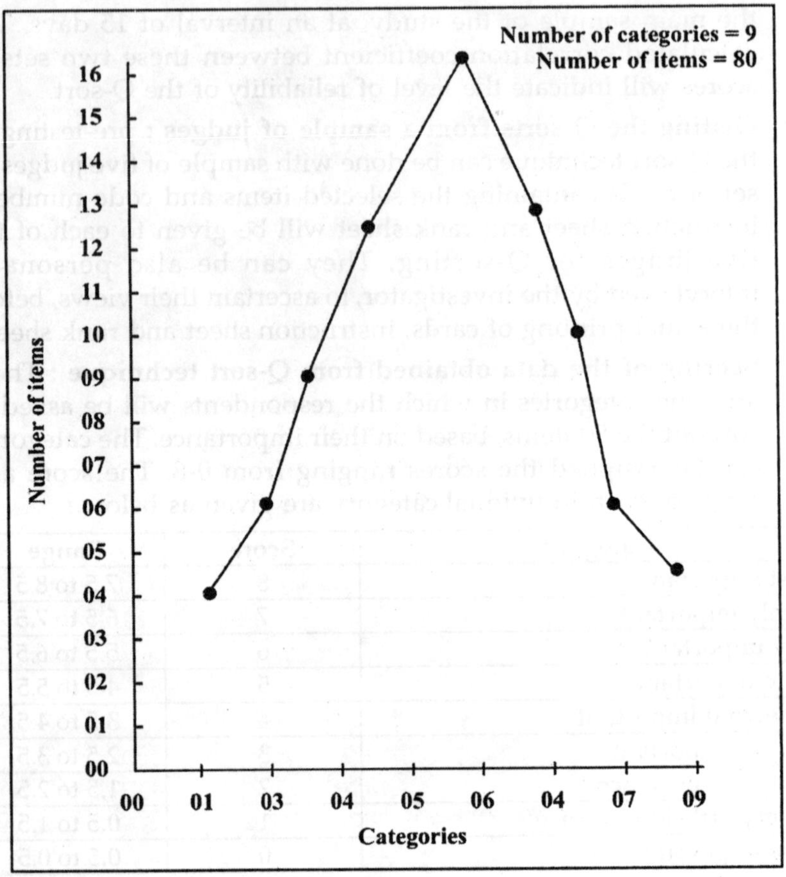

REFERENCES

1. Block, J. (1952). The Q-sort in Assessment and some problems in its use. I. P. A. R. Research Bulletin.

2. Block, J. (1961). The Q-sort Method in personality Assessment and psychiatric Research. Illinois : Charles C. Thomas publisher, Spring field.

3. Cattell, R.B. (1944). psychological Measurement : Ipsative, Normative, and Interactive. Psychological Review, 51 : 292-303.

4. Girdar, P.N. (1968). A Study of Factors Resisting the progress of Artificial Insemination in Haryana, M.Sc. Thesis. Haryana Agricultural University, Hissar.

5. Jaiswal, N.K. (1965). A Study of Factors Associated High, Low Level of Adoption of Improved Agricultural practices, Ph. D. Thesis. Division

of Agricultural Extension, I.A.R.I, New Delhi.

6. Jalihal, K. A. (1970). A study on the Concept and Role of Agricultural Universities in India, Ph. D. Thesis. Division of Agricultural Extension, I. A. R. I, New Delhi.

7. Jha. P. N. (1968). A critical Analysis of Factors Associated with communication Fidelity with Reference to High Yielding Varieties Programme. Ph.D. Thesis. Division of Agricultural Extension. I. A. R. I., New Delhi.

8. Kerlinger, Fred N. (1964). Foundations of Behavioural Research. New York : Halt, Rinehart, and Winston.

9. Livson, N. H. and Nichols, T. F. (1856). Discrimination and Reliability in Q-sort personality Descriptions. Journal of Abnormal and Social Psychology, 52 : 159-65.

10. Mac Kinnon, D. W. (1958). An Assessment study of Air Force Officers ; part I, Design of the study and Description of the Variables. Texas : personal Laboratory, wright Air Development Centre (Technical Report WADC-TN 58-91. Part I : ISTIA, Document No. A. D. 152840).

11. Marsh, Shirley, A. (1963). Q-technique : An Advance in Idio-dynamics. Paper issued by the Communication Research Centre, College of Communication Arts, East Lansing. Michigan State University, Michigan, U. S. A.

12. Nunnaly, J. (1967). Psychometric Theory. New York : McGraw Hill Book Co. Ltd.

13. Ramesh Babu, A. and Singh, Y. P. (1984). Pusa Rank sheet for Q-sort Technique. Indian Journal of Extension Education. Vol. XX (172) : 52-57.

14. Sontag, M. (1968). Attitudes Toward Education and perception of Teacher Behaviours. American Educational Research Journal, 5 : 385-402.

15. Stephenson, W. (1936) - Introduction to Inverted Factor Analysis, with some Applications to studies in Grexis, Journal of Educational psychology. 27 : 353-67.

16. Stephenson, W. (1953). The study of Behaviour, Chicago : University of Chicago Press.

Chapter **32**

Pragmatic Approach to Evaluation of Impact of Communication Programme

D. Das Gupta

Sorting out the good from the bad has long been a human activity, and one that is essential to society itself. Although institutions have often been accorded authority for making certain value judgments, only in recent times has evaluation of public programmes become so formalized as to constitute a recognizably separate activity - and even a new discipline.

Evaluation leads to settled opinion that something is the case. Evaluation leads to a judgment about the worth of something. This judgment often is arrive at by gardening or by making something according to how well the object fulfils a set of standards or criteria. It is essentially comparative, and usually explicitly so. This means there must be a set of standards and a class to which the object is compared. So, in its essence, evaluation entails adopting a set of standards, defining the standards, specifying the class of comparison, and deducing the degree to which the object meets the standards.

Models/approaches :

There are many advocated approaches to evaluation. Yet most of the approach can be subsummed under a few basic types. Some call these basic types "models" to indicate that they are designs or structural types advocated by prominent theorists as approaches worth imitating.

The taxonomy of the major approaches/models of evaluation is given in the Table-1. In the first column of the taxonomy, the common names of the approaches are listed. In the top row, are the critical dimensions of comparison : the audiences to whom the evaluation is addressed, which approaches assumes consensus on, methodology of data collection, the ultimate outcome expected and the typical questions that the approach tries to address. At one level of abstraction, these are the assumptions underlying the approaches.

Table-1 : Taxonomy of major approaches/models of evaluation

Model	Major audiences or reference groups	Assumes consensus on	Methodology	Outcome	Typical questions
1	2	3	4	5	6
System analysis	Economists, managers	Goals, known cause and effect quantified variables	PPBS, liner programming, planned variation, cost-benefit analysis	Efficiency	Are the expected effects achieved? Can the effects be achieved more economically? What are the most effective programmes?
Behavioural objectives	Managers, psychologists	Prespecified objectives, qualified outcome variables	Behavioural objectives achievement tests	Productivity, accountability	Is the programme achieving the objectives? Is the programme producing results?
Decision making	Decision makers, specially administrators	General goals, criteria	Survey, questionnaires, interviews	Effectiveness, quality control	Is the programme effective? What parts are effective?
Goal target.	Consumers	Consequences criteria	Bias control, logical analysis, modus operandi	Consumer choice, social utility	What are all the effects?
Art criticism	Connoisseurs, consumers	Critics standards	Critical review	Improved standards heightened awareness	Would a critic approve this programme? Is the audience appreciation increased? How would professional rate this programme?
Professional	Professionals, public	Criteria, panel, procedures	Review by panel, self study	Professional acceptance	How would professional rate this programme?
Quasilegal	Jury	Procedures and judges	Quasi legal procedures	Resolution	What are the arguments for and against the programme?
Case study	Client practitioners	Negotiations, activities	Case studies, interviews, observations	Understandin g diversities	What does the programme look like to different people?

Note : PPBS : Planning, Programming and Budgeting System.

In the taxonomy, the models are related to each other in a systematic way. Generally, the more one progresses down the column of major audiences, the more democratic or less elitist the audience becomes. The more one moves down the consensus column, the less consensus is assumed on goals and other elements. The more one moves down the methodology column, the more subjectivist and less objectivist the research methodology becomes. The more one moves down the outcomes column, the less the overall concern becomes for social efficiency and the more it becomes personal understanding. These are over implications since the actual ordering is more complex.

Forms/types of evaluation :

Forms/types of evaluation are basically governed by the degree of evaluation which are again to be illustrated by means of continuum.

Table-2 : Forms/types of evaluation

Forms/ types of evaluation	Major features	Degree of evaluation	Remarks formal/informal
1	2	3	4
Casual every day evalution	Based on first hand impression (s) of a particular stimulus, and not on principle (s) of evaluation. e.g. quality of simple information	Very low	Highly informal
Self checking evaluation	Based on certain principle (s) and conscious self-checking	Low to very low	Informal
Do it yourself evaluation	Systematic "doing" along with "checking" is needed; "doing" may be mentally and/or physically; involves planning and application of principles of evaluation e.g. judging of a journal by score card/content analysis.	Moderate	Formal
Survey/extension studies	Much systematic; based on standard tools & techniques e.g. thesis, through investigation, case studies, etc.	High	Highly formal
Scientific research	Involves standard research method; sometime may involve a team of specialists/scientists	Very high	Extremely formal

Measuring devices/tools :

There are various types of measuring devices for obvious reason of getting/evidences as to one's skill, attitude, knowledge, interest, adoption and so on. Such may be either qualitative or quantitative. Few important devices for measuring the changed behaviours/impact are stated below :

Table-3 : Measuring devices and characteristics functions

S. No.	Type	Functions/characteristics
1.	Value scale	Determination value of object/stimulus; rates the importance of thing/condition/circumstances.
2.	Attitude scale	Determines how one feels towards certain object/problem/programme/stimulus; ascertains agreement or disagreement (with degree) towards certain fact/problem/issue/stimulus
3.	Opinion polls	Ascertain the opinion of people on some stimulus/issue
4.	Knowledge/ comprehension test	Records the knowledge level of clients as a test of programme; or finds out whether one understand and/or can apply certain knowledge when need arises.
5.	Skill performance rating	Assesses the amount of skill required by one as a result of a programme like training
6.	Interest rating or checking	Determine the extent of interest one has about some programme/activity/method
7.	Adoption rating	Ascertains the degree/extent of adoption of practices
8.	Case studies	Studies in depth only one or few units of object/institution/programme etc. Sometimes genesis of the development of a project/activity is studied in detail and reported.

While constructing a device for measurement or in applying it, one should check the following criteria very carefully.

(a) Reliability

(b) Validity

a. Behaviour validity

b. Content validity/subject matter validity

(c) Objectivity

(d) Practicability

(e) Simplicity

Five criteria with their mutual relationship may be explained by the model as–

Reliability
Validity
Objectivity
Practicability
Simplicity

Δ Ability to give consider results.

° Ability to assess character in correct form.

° Gets evidence of changes behaviour (behaviour validity).

° Deals with subject matter/content.

Δ Ability to study without prejudice/bias/impression.

Δ Pragmatic & applicable considering factors of cost time (resource) convenience.

Δ Simple and easy to understand, answer, analyze and summaries.

Experimental designs :

If we want to test the hypothesis that 'A' is a cause of 'B' by comparing a group that has been exposed by 'A' with one that has not, it is essential to measure the two groups with respect to 'B' either during or after exposure/experiment to 'A'. It is also essential to have measure of their position with respect to 'B' before they have been exposed to 'A'. On the basis of the time at which the dependent variable is measured provides a basis for classifying experiments/exposure in two main groupings : (a) "AFTER-ONLY" and (b) "BEFORE-AFTER". Moreover, the pattern of control groups used gives a basis for further subdivision (see table 4 &5).

Major dimensions of study in communication research :

With the valid and objective criteria study and various designs of experiments in possession one may get into the investigation of proper dimensions in research in communication/technology transfer. It is impossible to make an exhaustive list of such dimensions. However, some of the major dimensions are as follows :

(a) Adoption and/or retention of advocated technologies.

(b) Technology dissemination.

(c) Change/shift in
   a. Behaviour-overt and covert
   b. Skill and ability
   c. Value
   d. Interest
   e. Attitude
   f. Knowledge and comprehension

(d) change/shift in
   a. gross coverage under specific programme/stimuli
   b. cropping pattern and cropping intensity
   c. overall land use pattern
   d. production and productivity
   e. income and its distribution
   f. employment pattern
   g. asset accumulation

(e) Factors of adoption/retention and rejection of advocated technologies.

(f) Reasons of non-adoption/rejection of advocated technologies.

(g) Linkage and feedback

(h) Constraints analysis

Table-4A : Types of experimental design

Conditions	"AFTER ONLY"		"BEFORE-AFTER" with single group/experimental group
	Experimental group	Control group	
Prior selection of groups	Yes	Yes	Yes
'Before' measurement	No	No	Yes (B_1)
Exposure to experimental variables	Yes	No	Yes
Exposure to uncontrolled event	Yes	Yes	Yes
'After' measurement	Yes (B_2)	Yes (${B^1}_1$)	Yes (B_2)
Change/impact	$d = {B^1}_2.B - B_1$		$d = B_2 - B_1$

Table-4B : Types of experimental design

Conditions	"BEFORE AFTER' with one control group		"BEFORE-AFTER" with two control groups		Control group II
	Experi-mental group	Control group	Experi-mental group	Control group I	
Prior selection of groups	Yes	Yes	Yes	Yes	Yes
'Before' measurement	Yes (B_1)	Yes ($B^1{}_1$)	Yes (B_1)	Yes ($B^1{}_1$)	No $B^{1} = \frac{B^{1} + B_{1}^{1}}{{}^{2}\text{Yes}}$
Exposure to experimental variables	Yes	No	Yes	No	Yes
Exposure to uncontrolled events	Yes	Yes	Yes	No	Yes
'After' measurement	Yes (B_2)	Yes ($B^1{}_2$)	Yes (B_2)	Yes ($B^1{}_2$)	Yes (B_2)
Change/impact	$d = B_2 - B_1$	$d = (B^1{}_2 - B^1{}_1)$	$d = B_2 - B_1$	$d^1 = B^1{}_2 - B^1{}_1$	$d = B^1{}_2 - B^1{}_1$

Table-5 : Salient aspects of major experimental designs

"AFTER ONLY' design	"BEFORE-AFTER" with single group	BEFORE AFTER with one control group	BEFORE-AFTER with two control groups
1	**2**	**3**	**4**
Only after exposure (in some cases during the exposure) measurements are made	Each subject serves as his own control	It is employed for getting the effectiveness of the exposure alone	It can separate the influence of the exposure variable from that of the initial measurement even if there is interaction between them
One concludes that the hypothesis is or is not tenable by comparing the occurrence of 'A'	Its use is justified only when one has good reason to be -lieve (a) that the 'before' measure itself will not in	It provides adequate safeguards against error in the measurement procedure	

"AFTER ONLY' design	"BEFORE-AFTER" with single group	BEFORE AFTER with one control group	BEFORE-AFTER with two control groups
1	2	3	4
in the experimental group after exposure to 'A', with the occurrence of 'B' in the control group	some way affect either the response to the experimental treatment or the 'after' measure; and 'b' that there are not likely to be any other influence besides the exposure during the course of study		
Provides proof as to whether 'A' and 'B' vary together		It fails to provide any way of discovering the interaction and separating out the effect of the exposure variable alone	
It is less serious than "Before-after design			

Important indices and statistical tests applicable for communication research :

Communication researchers have to take the help of various indices and statistical tests in order to draw fruitful conclusions. Some most important indices and statistical techniques are given below :

1. Indices :

(a) Cropping intensity

$$C.I._j = \frac{Mi}{Ai} X\ 100$$

Where M_i = gross cultivated land in as/ha ith farm,

A_i = net cultivated land in acre in ith farm and

$C.I._i$ = cropping intensity in ith farm

(b) Productivity index

$$P.I._i = \frac{\sum_{j=1}^{k} \frac{Y_{ij}}{Y_i} X\, m_{ij} X\, 100}{A_i}$$

Where Y_{ij} = Yield per ac/ha in quintal of jth crop in the ith farm

Yj = mean per ac/ha in quintal

Mij = area in acre/ha under jth crop in ith farm

Ai = net cropped land in ac/ha in ith farm

i = 1, ... n (n being the number of farm family under study

j = 1, ... k (k being the number of crop under study.

Crop yield index :

Tambad (1965) and Tambad and Patel (1970) have adopted 'Crop Yield Index' as the basis for measuring agricultural productivity. The purpose of this technique is to express the average yield of vatious crops on a farm or in a region relative to the yield of same crops on another farm or in a second region. It can be expressed in the equation as –

Bhatia (1967) while assessing the changes and trends in agricultural efficiency adopted the measurement of productivity index as –

$$\text{Crop yield Index} = \frac{\left[\sum_{t=1}^{n} \frac{Y_i}{Y_{io}}\right] A}{\sum_{i=1}^{n} A_i}$$

Where I = 1, 2, 3, ...n are the number of crops considered in an unit area of year.

Y_i = is the yield/ac of crop i, in a farm area or year,

A_i = is the weightage of crop i, denoted by the area under crop as a percentage of total cropped area, and

Y_{io} = is the average yield/ac of crop i, at the group of farms

Bhatia (1967) while assessing the changes and trends in agricultural efficiency adopted the measurement of productivity index as –

$I_{ya} = [Yc/Yr] \times 100$

Where I_{ya} = the yield index of crop 'a'

Yc = the average acre yield of crop 'a' in the component unit and

Yr = the average yield/acre of crop 'a' in the entire study/area.

Income distribution :

The procedure towards an estimation of the income distribution. Lorenz Curve as has been put forward by Kakwani and Poddar (1976) as NEW COORDINATE SYSTEM FOR THE LORENZ CURVE. This process has wide empirical validity. The well-known measures of inequality may be obtained as the function of the estimated parameters of the Lorenz Curve. Moreover, the frequency distribution can also be derived from the equation of the Lorenz Curve. The estimated curve can be used to obtain any percentage of the distribution.

The income inequality is measure in terms of Gini Concentration Ratio (GCR) computed with the help of New Coordinate System for the Lorenz Curve.

Statistical Tests :

Analysis of variance ('F' test) :

The analysis of variance of 'F' furnishes comprehensive or overall test of the significance of the difference between means. A significant 'F' do not tell us which mean values differ significantly, but that at least one is reliably different from some others. If 'F' is not significant, it is not essential for further testing, as none of the mean differences will be significant. But if 'F' is significant, it is essential to test the separate differences by the 't' test. This 't' can be computed by using the formula :

$$f = \sqrt{F}$$

or, for testing the null hypothesis Ho ($u = u_0$) i.e., from two normal populations with menu u_1 and u_2 from two sample sizes (standard deviation is unknown) n_1 and n_2 't' test is applied.

$t = (X_1 - X_2 \text{ , } S \sqrt{(1/n_1 = 1/n_2)}$ [x_1 and x_2 are sample means]

(known as Fisher's 't') follows 't' distribution

$[(X_1 - X_2) - (U_1 - U_2) \text{ , } S \sqrt{(1/n_1 = 1/n_2)}]$ with

$(n_1 + n_2 - 2)$ degrees of freedom.

Where,

$S = v[(n_1 s^2_1 + n_2 s^2_2) . (n_1 + n_2 - 2)]$

(s_1 and s_2 are the sample standard deviations).

The phi co-efficient (Ø) :

When the classification is truly discrete and the variables can take only one of two values, the Ø coefficient is an appropriate measure of correlation. The formula is Ø is :

$$Ø = \frac{AD\text{-}BC}{[(A+B)(C+D)(A+C)(B+D)]}$$

Where Ø = co-efficient of correlation. A, B, C and D represent frequencies in four fold (2 x 2) table (Garrett and wood worth).

In the present study, Ø is used to determine the item-item correlations. The significance of Ø may be determined through the relationship of Ø to x^2; x^2 is related to Ø by the following equation :

$X^2 = NO_2$ (Garrett and woodworth)

Where, N is the number of pairs under study,

And X^2 as a function of Ø.

Or,

$$X^2 = \frac{N(AD\text{-}BC)^2}{R_1 R_2 C_1 C_2}$$

Where, R^1 R^2 are Rows and C^1 C^2 are columns respectively, if observed, X^2 value is greater than the tabulated value, the null hypothesis is rejected and the alternative hypothesis is accepted.

Yates' correction :

For a (2 x 2) table, there is only one degree of freedom; i.e. only one of the four cell frequencies can be arbitrarily given, the row and column totals should remain fixed. It is therefore, necessary to make a correction to the above X^2 formula so that its approximation to the continuous chi-square distribution can be improved.

$$X^2 \text{ (corrected)} = \frac{N[AD\text{-}BC\ [-N/2]^2}{R_1 R_2 C_1 C_2}$$

with 1 degree of freedom.

Multiple correlation :

A multiple correlation coefficient measures the combined relation between a dependent and series of independent variables. It can also be explained as the correlation between the observed values of the dependent variable and its estimated values from the independent variate values, estimated with the help of the multiple regression equation.

If Y is the dependent variable and X_1 X_2 ..... are the independent variables, then the multiple regression equation will be

$y = a + b_1 x_1 + b_2 x_2 \ldots$

where, a is a constant and b_1 b_2... are the partial regression coefficients.

Multiple regression analysis :

The main use of a regression equation is to predict the most likely measurement in one variable from the known measurement in another. In order to apply this type of analysis, here it is assumed that a linear relationship exists between a variable y and k-1 explanatory variables x_2 x_3 ... x_k and a disturbance term u. If a sampled of n observation on y and the x's is there one can write

$Y_j = b_1 + b_{2\ x\ 2i} + \ldots b_k x_{ki} + u_i \ldots$ (1)

$i = 1, 2, \ldots\ldots\ldots\ldots\ldots\ldots\ldots\ldots$ n

The b coefficients and the parameters of the u distribution are unknown and the problem is to obtain their values. The n equations in (1) can be set out compactly in matrix notation as

$Y = Xb = u$ .... (2)

$$\text{Where } Y = \begin{bmatrix} Y_1 \\ Y_2 \\ . \\ . \\ Y_n \end{bmatrix} \quad X = \begin{bmatrix} 1\ X_{21} \ldots\ldots\ldots\ldots X_{k1} \\ 1\ X_{22} \ldots\ldots\ldots\ldots X_{k2} \\ . \\ . \\ 1\ X_{2n} \ldots\ldots\ldots\ldots X_{kn} \end{bmatrix}$$

$$b = \begin{bmatrix} b_1 \\ b_2 \\ . \\ . \\ b_k \end{bmatrix} \quad u = \begin{bmatrix} u_1 \\ u_2 \\ : \\ . \\ u_k \end{bmatrix}$$

Under the assumptions (i) ui are variables with zero expectation and u_i values are pairwise uncorrelated, (ii) X is a set of fixed numbers and (iii) X has rank k. n. The least-squares principle is applied to estimate the parameters of (1) Let

$b = b_1 b_2 \ldots\ldots b_k$ .... (3)

denote 'a' column vector of estimates of b, then it can be written

$Y = Xb + e$ .... (4)

Where 'e' denotes the column vector of n residual (y-xb) To find the value of b which minimizes the sum of squared n

Residuals e_i 2 is differentiated and equating it to zero the result obtained

$X' X. b = X' Y$

Where X' is the transport matrix of X. This expression can be written as

$b_{-1} = (X'X)^{-1} X'Y$

Where (X'X) is the inverse matrix of (X'X).

This is the fundamental result for the least-square principles.

Path analysis :

Path analysis adopting the multivariate Path Model as suggested by Land (1969) is used to isolate the direct as well as indirect effect of exogenous variables. Obviously, the general principles of path analysis are more attuned to construction of unidimensional causal theories, but there are ample evidences where it has proved its efficiency in handling reciprocal relationship as well. From the epistemological and scientific logic perspective, we concur that path analysis is a better tool of analysis than simple correlation and regression models.

The cause and the effect relationship is well defined, it is possible to represent the whole system of variables in the form of a diagram, known as path-diagram. Let us consider that the yield 'Y' of barley is the function (effect) of various components (causal factors) like number of ears per plant (X_1), number of grains per ear (X_2) and weight of 100 grain (X_3) etc. Let us also assume that these factors show the following type of association with each other (Diagram-a)

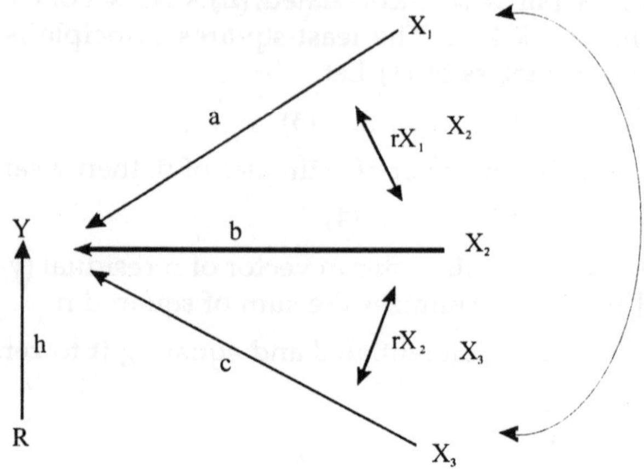

Cause and effect relationship

From the above figure, it is obvious that yield is the result of X_1, X_2 and X_3 and some other undefined factors designated by R. Further, X_1, X_2 and X_3 in turn are correlated. In the diagrams a and b are the path coefficients due to respective variables. Path coefficient can be defined as the ratio of the standard deviation of the effect due to given cause to the total standard deviation of the effect i.e., if y is the effect and x_1 is the effect y is $ox_{1/0y}$.

To make it more explicit, let us consider a simple system of independent (x1 and x2) and dependent (y) variables (Diagram b). The interrelationship as shown in this figure may be expressed mathematically as :

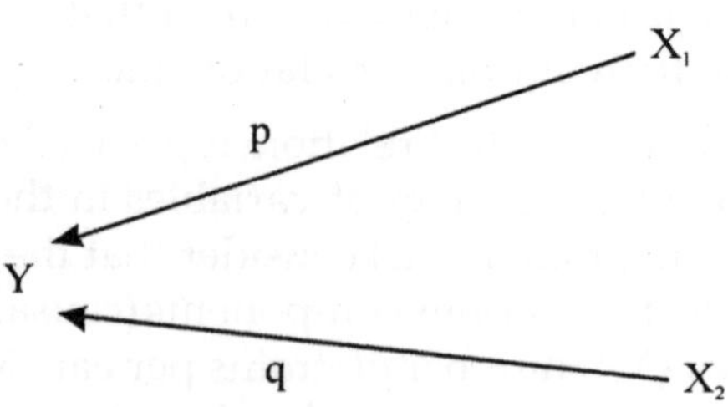

Path Diagram

$Y = X_1 + X_2$

Or

$O^2Y = O^2X^1 + O^2X^2$

(Assuming that x_1 and x_2 are independent)

Let us divide both sides of the above equation by O^2y to get :

$O^2Y \div O^2y = (O^2y \div O^2y) + (O^2x_2 \div O^2y)$

Or, $(O^2x_1 \div O^2y + (O^2x_2 \div O^2y) = 1$

Denoting $p = ox_1 \div o_y$ $q = ox_2 \div o_y$

Here p and q are called 'path co-efficient' and p^2 is the co-efficient of determination.

The advantage of the path diagram is that a set of simultaneous equations can be written directly from the diagram and a solution of these equations provides information on the direct and indirect contribution of these causal factors to the effect.

Stepwise multiple regression analysis :

According to Draper and Smith, the method of stepwise multiple regression analysis is to insert variables in turn until the regression equation is satisfactory. The order of insertion is determined by using the partial correlation coefficient as a measure of the importance of variables not yet in the equation.

The programme, according to Burrough Corporation, first forms a correlation matrix, finds the best predictor (the independent variable with the second highest correlation with criterion) is found and a regression analysis using the multiple correlations of the two predictors is performed. At any given step, the group of that size (i.e., the particular group of independent variables does not necessarily have the highest multiple correlation with the criterion that any group of this size does). Rather this group contains the variables that have the highest individual correlations with the criterion.

Significance of a variable that is being considered for entrance into the regression equation is measured by the F statistic. If f is too small (less than F 'include'), the variable is not added to the regression equation. 'Include' statement establishes the minimum value of the F-statistics required for the inclusion of a variable in the regression equation. For the purpose of the present study F-value for inclusion was 0.01.

Significance of variables already in the regression equation may change as new variables are entered. The significance of the variables currently in the equation is also measured by the F. statistics. If F is too small (less than F 'delete') the variable is not added to the equation. 'Delete' establish the value of the F-statistics below which the variable is deleted from the regression equation. Here the F-value for deletion was 0.005.

The 'tolerance' level specified is used as control of degeneracy. Degeneracy occurs when a variable entered into the equation is a linear combination of variables already present in the equation. 'Tolerance' statement establishes the maximum value a pivoted element may attain while still allowing its associated variable to be brought into equation. A variable is not brought into the regression equation if its associated pivoted element is below the specified tolerance level.

Factor analysis :

The principal, concern of factor analysis, according to Harman, is the resolution of a set of variables linearly in terms of (usually) a small number of categories of 'factors'. This resolution can be accomplished by the analysis of correlations among the variables. A satisfactory solution will yield factors which convey all the essential information of the original set of variables. Thus, the chief aim is to attain scientific parsimony or economy of description.

The original variable X_j is represented by the following linear composite expressed in terms of the factor as follows :

$$X_j = \sum_{r=1}^{P} 1_{jr} \quad f_r$$

Where p is the number of factors retained and f_r (r = 1, 2, p) represents common factors and 1_{jr} are the factor loading of the original variable X_j on factor f_r.

The method of principal component analysis is followed in the present study. The principal component analysis extracts m-eigenvectors (Principal component axes) and corresponding m-eigenvalues, (the variance measure along the eigenvector) from m x m symmetrical matrix of correlation. The eigenvectors obtained from this Principal Component analysis are all orthogonal (i.e., intercolumn correlations are near zero). The eigenvalues account for all of the original data variances in decreasing order such that each has variance or eigenvalue less than the previous ones. The total of the eigenvalues $\wedge_1 + \wedge_2 + \ldots + \wedge_m$ which is the same as the sum of the variances constituting the diagonal or trace of the correlation matrix before transformation. The principal components are then converted into factors by multiplying each element of the principal components or eigenvectors (V) by the squars-root of the corresponding eigenvalues (^1/2 V) factors, thus, besides the direction also represent the variances.

The analysis calls for the selection of a minimum number of meaningful and useful factors, considerably fewer in number ($p < m$) than the original variables, which will account for most of the variances in the data set and therefore, convey the same information. Various criteria selection of suitable factors are available. Kaiser and others have recommended retaining all those eigenvalues which have values more than 1. This criterion was followed in the present study.

Next step is to remove the noise imposed by (m-p) unnecessary axes. To accomplish this, p-orthogonal reference axes or factors are rotated about the origin to positions such that the variance of the loading from each variable onto each factor axis is either extreme (+1) or near zero. This maximization of the range of the loading is performed by using Kaiser's varimax criterion. Scanning through each factor column for large absolute values in the varimax matrix will reveal a few variables with significantly high loading and many others with insignificant loadings. The column showing communitality ($?h^i_2$) is the total amount of variance of each variable retained in the factors, and is computed by summing the squares of the elements of the factors in each row of the varimax matrix. Fairly high communality of each variable implies the appropriateness of the model adopted.

Choosing an appropriate statistical test :

Normally, there are alternative and valid statistical tests for a research hypothesis. In such situation, there should be some rationale for choosing among them. Actually, there are some considerations for choosing a particular statistical test. In this choice, one must consider - the manner in which the sample or scores/data is drawn;

- The nature of the population from which the sample is drawn;
- The particular hypothesis one to test; and
- The kind of measurement/scaling which is employed in the operational definitions of the variable involved i.e., in the scores.

All these enter into determining which test is optimum or most appropriate for analyzing a particular set of research data.

Measurement and its levels :

Scale	Explanation (s)	Defining relations
Nominal/categorical/ classificatory	When numbers/other symbols are used simply to identify and classify an object, person, characteristics, groups example : neurotic, manic depression	Equivalence
Ordinal ranking	Given a group of equivalence classes (i.e., given a nominal scale), if the relation holds between some but not all pair of classes, one have a partially ordinal scale. If the relation holds for all pairs of classes so that a complete rank ordering or classes is possible, one have an ordinal scale. Example : Socio-economic status	Equivalence Greater than
Internal	If a scale has all the features of an ordinal scale, and if the addition the distances/differences between any two number on the scale having meaning, then measurement considerably stronger than ordinal has been achieved. In such a case measurement has been achieved in the sense of an interval scale. Example : Temperature measurement	Equivalence Greater than Known ratio of two intervals
Ratio	When a scale has all features of interval scale and in addition has a true Zero Point as its origin, it is called a ratio scale. Example : Mass or weight	Equivalence Greater than Known ratio of two intervals Known ratio of any two scale values

Nominal and ordinal measurements are the most common types achieved in the social and behavioural sciences. Data measured by either nominal or ordinal scales must be analysed by non-parametric methods. On the other hand, data measured in interval or ratio scales may be analysed by parametric method, if the statistical model is valid for the data almost all the exact (small) sample tests of significance are based on the fundamental assumption that the parent population is normal and are concerned with testing or estimating the means and variance of these populations. Such tests, which deal with parameters of the population are known as Parametric Tests, a parametric statistical test is a test whose model specifies certain conditions about the parameters of the population from which the samples are drawn.

On the other hand, a Non-paramatric test is that does not depend on the particular form of the basic frequency function from which the samples are drawn. In other words, non-parametric test does not make any assumption regarding the form of the population.

In short, the criteria for selection and use of a statistical test for use in making a decision about a research hypothesis are (a) the applicability or validity of the test (which includes the level of measurement and the other assumptions of the test) and (b) the power and efficiency of the test.

Suggested books and journals :

1. Anonymous (1973). Burroughs advanced statistical inequity system user's manual. Burroughs corporation, Detroit, Michigan, USA.
2. Bailey, D.E. (1971). Probability and statistics : models for research. New York : J. Wiley.
3. Bhatia, S. S. (1974) Spetial variations changes and trends in agricultural efficiency U.P., 1953-56. Indian Journal of Agricultural Economics, 22 (1) pp. 66-20.
4. Chandel, R. S. (1970) A handbook of agricultural statistics. Achal prakashan mandir, Parmat, Kanpur.
5. Davidson, D., suppes, P., and Siegel, S. (1987). (Decision making : an experimental approach. Stanford, CA : Stanford, University Press.
6. Draper, N. R. and Smilt, H. (1981) Applied regression analysis John Wiley and Sons Inc., New York.
7. Frutchey, F. P. (1967) Evaluation in Extension, Division of Extension research and training, Federal Extension Service, USDA.
8. Garrett, H. E. and R. S. Woodworth (1971) Statistics in Psychology

and Education (6th edition), Vakils, Fefter and simons Pvt. Ltd. Bombay, pp. 127-79 an 389-91 (Original publisher David Mckay Company Inc., New York).

9. Goode, W. J., and Hatt, P.K. (1952) Methods in social research Mcgraw Hill.
10. Gupta, S. C. and Kapoor, V. K. (1989) Fundamentals of mathematical statistics sultanchand and sons publishers, New Delhi.
11. Haman Harry, H. (1976) Moden factor analysis. The Unviersity of Chicago.
12. Hays, W. L. (1960) a note on average tax as a measure of concordance. Jr. of Amer State Assoc. 55. 331-341.
13. Hays, W. L. (1981) Statistics (Third edition), New York : Holt. Rinehart and Winston.
14. House, E. R. (1990) Evaluating with Validity Sage Publicatons, London.
15. Kaiser, H. F. (1958) The Varimax Criterion for analytic rotation in factor analytic rotation in factor analysis, Psychometrika. 23, pp. 187-200.
16. Kakwani, N. C. and Poddar, N. (1976) Efficient estimation of the Lorenz Curve and associated inequality measures from grouped bservations. Econometrica 44(1), pp. 137-148.
17. Kaplan, A. (1943) Content analysis and the history of signs. Philosophy of Science, 10, 247-250.
18. Land K. C. (1959) Principals of path analysis,Edger F. Borget (Ed., Sociological methodology, Jessey Ban. Sanfransisco, pp. 3-7.
19. Lzersfeld, P. F. (1957) Latent structure analysis, Bureau of applied social research, Columbai University, Mimeographed Rutman, L; 1980, Planning useful evaluation : Evaluability assessment, Carleton University.
20. Seltiz, C., Jehoda, M. and Cook, S. W. (1959). Research methods in social sciences. Plelhuen and Co. Ltd., USA.
21. Siegel, S. and Castellan, N. John, Jr. (1988). Nonparametric statistics for the behavioural sciences (second edition) McGraw Hill.
22. Tambad, S. B. (1965) Social and temporal variation in agricultural productivity in Mysore Indian Journal of Agricultural Economics, 20, p. 41.
23. Tambad, S. B. and Patel, K. V. (1970) Crop yield index as a measure of productivity. Economic and political weekly. 5 (25). Pp. 878-879.
24. Taylor, P. W. (1961) Notmative discourse. Englewood cliffs, NJ : Prentice Hall.
25. Townsed J. T. and Ashby, F. G. (1984) Measurement scales and statistics : The misconception miscoriceived. Psychological Bulletin. 96, 396-401.

Chapter **33**

Modeling Innovation Diffusion – A Review

Sibaji Chakraborti

Diffusion is a communication process of an innovation which passes through certain channels over time among the members of a social system. In General, the process initiates only with a few members of the system adopting the innovation, subsequently an increasing number of adoptions take place per unit of time and finally reaches an asymptote when the diffusion is complete. Thus,3 the cumulative adoption time path or temporal pattern of diffusion process is a sigmoid (S-shaped) curve.

Numerous hypotheses and interpretations have been offered by various researchers regarding sigmoid nature of the diffusion curve. These are economic advantage based explanation (Manas-field 1961), or demand and supply based explanation (Griliches, 1957, Brown 1981), or learning perspective based explanation (Casetti and Semple, 1969, Sahal, 1981), or information transfer based explanation (Hargerstrant 1967, Bernhardt and Mackenzie 1972), or technology substitution based explanation (Blackman 1974, Sharif and Kabir 1976). Finally, came the communication (Regel, 1983) for interpreting the diffusion process.

Much of the early researchers on diffusion process used prespecified trend or distribution functions which give rise to S-shaped curves, like cumulative normal, Logistic or Gompartz curve. However, since any unimodal distribution functions will give rise to S-shaped curve, it is often impossible to find the best fitted curve empirically. Therefore, attempts were made to develop theory-based diffusion models.

The fundamental diffusion model can be written as a differential equation.

$$\frac{dN(t)}{ct} = g(t) \left[\bar{N} - N(t)\right] \text{.................................} \quad (1)$$

With $N = N_0$ for $t = t_0$

Where N (t) = Cumulative number of adopters at time t

$\bar{N}$ = Total number of potential adopters at time t

g (t) = Coefficient of diffusion at time t

$\frac{dN(t)}{dt}$ = Rate of diffusion at time t.

This diffusion model is a deterministic rate equation which states that the rate of diffusion at any point of time t is a function of [$\bar{N}$-N (t), that is, difference between the total number of potential adopters and actual number of adopters already adopted at the beginning of time t.

The nature of relationship of this model is controlled by g(t). The coefficient of diffusion. It depends on the nature of innovation, communication channels employed, and social system attributes. This g(t) can also be interpreted as the probability of adoption at time t. Then g(t) [$\bar{N}$ -N(t)] becomes the expected no. of adopters at time t, n(t). Thus, g(t) can also be viewed as a transfer mechanism, a conductivity coefficient or a conversion coefficient.

Two distinct approaches have been used to represent g(t) :

(1) g(g) is a function of time,

and

(2) g(t) is a function of previous number of adopters.

The second approach is obviously far more relevant and as such this approach will be discussed here. Specifically, g(t) can be expressed as a function of N(t) so that

$g(t) = a + b\,N(t) + cN^2(t) + \ldots\ldots\ldots\ldots$

However, for facilitation of interpretation and parametric estimation, g(t) is taken either

$g(t) = a$

or $g(t) = bN(t)$

or $g(t) = a + bN(t)$

Where 'a' and 'b' are model parameters.

When g(t) = a, the model is referred as the 'External-influence diffusion model, and the term 'a' is defined as index or coefficient of external influence coming from outside the social system, and in general represents the influence of change agent(s), on the diffusion process excluding prior adoption.

The External-influence diffusion model when integrated gives rise to a modified exponential curve, of the following figure 1. Diffusion processes with such a functional form are assumed only to be based on the information from an external communication source. Thus, this model assumes that the rate of diffusion at time t depends only on the potential number of adopters present at time t and that the model does not attribute any diffusion to interaction between prior adopters and potential adopters. In general, this kind of model is appropriate when members of a social system are isolated and do not interact.

tf g(t) = N(t) is the fundamental diffusion model (1). The model is known as the 'Internal influence' diffusion model. The model is thus -

$$\frac{dn(t)}{dt} = bN\ \bar{N} - N(t)] \text{ .... } (3)$$

Here, N(t)x [$\bar{N}$= N(t)] = (Prior adopters) x (Potential adopters), which is nothing but interpersonal communication or social interaction. Thus, the model is known as 'Internal-Influence' diffusion model which is treated solely as function of interpersonal communication. Hence, this model can be viewed as 'pure imitation' of diffusion model. The constant 'b; is defined as an index of imitation or internal influence as it reflects the interaction between N(t) and (N-N(t). Thus, 'b' can be viewed to represent the effect of horizontal channel (s) of communication, or decentralized channel (s) of communication, or unstructured informal channel (s) of communication.

The cumulative adopters' distribution function of this model is obtained by integrating eq. (3); corresponds to a logistic curve such as given below (figure II).

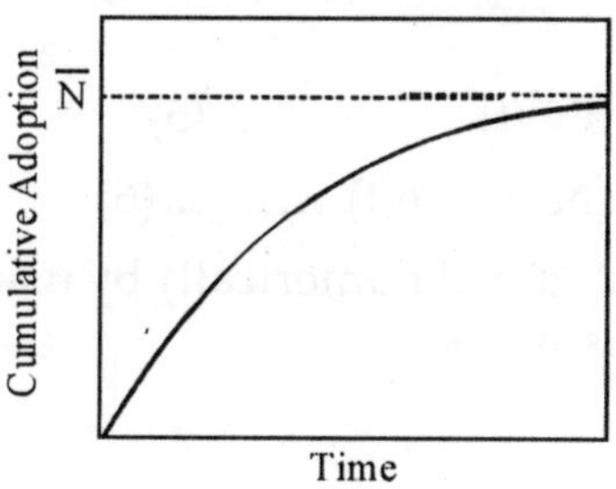

Fig. 1

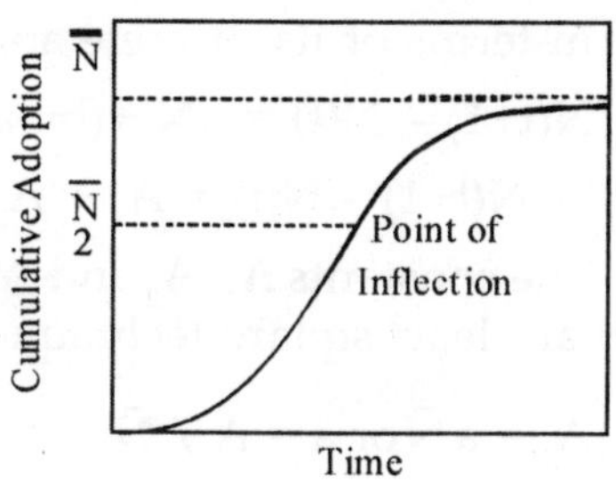

Fig. 2

The 'Internal-influence' diffusion model is directly related t Gompartz function, which is widely used in technological forecasting.

The internal-influence model is most appropriate when an innovation is complex and its adoption is socially advantageous. The social system in relatively small and there exists a need for verification of information prior to adoption.

When g(t) = a + bN(t) in equ. (1), we have a model which accommodates both external as well as internal influence and thus known as a 'Mixed-influence' model. The model then becomes.

$$\frac{dn(t)}{dt} = \left[a + bN(t)\right]\left[\bar{N} - N(t)\right] \quad \ldots\ldots\ldots\ldots\ldots\ldots\ldots \quad 4$$

This is the most general form of the three fundamental diffusion models.

Integration of the above model leads the cumulative adoption distribution results in a generalized logistic curve, the shape of which depends on both 'a' and 'b'.

Parameter estimation :

Estimation of parameters in the diffusion model requires time-series data. If such data are available, parameters can be estimated by standard and often nonlinear estimation procedure, or by using maximum likelihood estimation procedure. In ongoing diffusion process, initial parameters values can be revised as new data become available by using adaptive or Bayesiao estimation procedures.

A simple procedure for estimating the model parameters was suggested by Bass (1959). Let us consider the procedure with the general form of mixed-influence model :

$$\frac{dN(t)}{dt} = \left[a + bN(t)\right]\left[\bar{N} - N(t)\right]$$

in terms of its discrete analogue

$N(t+1) - N(t) = aN + (bn-a)\,N(t) - bN^2(t)$,............ (5)

or $N(t+1) - N(t) = A_1 + A_2N(t) + A_3N^2(t) + e(t)$ .... (6)

The constants A_1, A_2 and A_3 can be obtained numerically by using ordinary least square technique and thus we have

$A_1 = a\bar{N}$ or $a = A_1/\bar{N}$

$A_3 = -b$ or $b = -A_3$

And N as a root of the equation

$A_1 + A_2 N(t) + A_3 N^2(t) = 0$

$$\text{Which is } \bar{N} = \frac{-A_2 \pm A_2 \overline{_2} 4A_1 \quad A_3}{2A_3}$$

Observation on the fundamental diffusion models :

With the objective of facilitating analytical solution few simplified assumptions are made regarding the models. These are :

(1) Diffusion process is binary, that is, members of the social system either accept or reject it. Thus, it overlooks the stages of the diffusion process like awareness, knowledge etc.

(2) The social system will not change in size during the course of diffusion process. That is, the number of potential adopters (N) is a constant – either known or estimatable.

(3) There is only one adoption by an adopting unit and the model does not permit discontinuance of an innovation once it has been adopted.

(4) There is a complete mixing of prior and potential adopters with model parameters remaining unchanged over the diffusion process.

(5) Innovation itself does not change over the diffusion process and remains independent of all other innovation.

(6) Geographical boundaries of the social system remains unchanged over the entire diffusion process.

(7) The model is capable of accommodating all the remains unchanged over the entire diffusion process.

Let us now examine the basic structure of the fundamental diffusion model in terms of two mathematical properties :

(i) Point of inflection, where the rate of diffusion is maximum and (ii) symmetry. Both internal-influence and external-influence models have only limited flexibility in terms of these two properties. It can be shown that external influence model does not have a point of inflection, consequently, this model can not explain many diffusion patterns. In case of internal-influence model, it can be shown that, the maximum rate of diffusion is a constant and that occurs precisely when 50% of the potential adopters have adopted the innovation.

The internal-influence model is symmetric. This implies that the diffusion pattern after the point of inflection is the mirror image of the diffusion pattern before the point of inflection. It is obvious that this is not a logical proposition to assume. However, Gompartz curve of internal influence model though possesses a fixed point of inflection (for 37% adoption), is not symmetric.

The mixed-influence diffusion model, generates a symmetric diffusion curve with a point of inflection that must occur when 50% or below adopters have adopted the innovation.

Flexible models :

Because of lack of flexibility in the fundamental of diffusion models, several attempts have been made to develop more flexible model which would allow the generalized S-shaped diffusion curve to be either symmetrical or asymmetrical with the point of inflection responding to the diffusion pattern. These models can be obtained by means of the nonlinear or maximum likelihood procedure. Two such models are considered here as example. These models are due to Easingwood et al. (1981-1983). These are respectively known as the Non-symmotic responding Logistic (NSRL) model - a flexible internal - influence model, and the non-uniform influence (NUI) model, a flexible mixed-influence model. These models do not assume the consistency of the impact of internal influence between adopters and potential adopters over the entire diffusion process, and thus relaxes a questionable assumption made in the fundamental diffusion models. Easingowood et al. represented the impact of internal influence as a function of adoption level, a time varying effect, through

$W(t) = bF \alpha (t)$

Where α = constant; w(t) = Time-varying effect to internal influence and F(t) = N(t) = Fraction of potential adopters that has adopted the innovation by time t.

Hence, the NUI model is :

$$\frac{dF(t)}{dt} = \left[a + bF^{o}(t)\right] \left[1 - F(t)\right] \cdots\cdots\cdots\cdots \quad (7)$$

Where, o = 1 + α) = Non-uniform influence factor.

When α = o, that is, o = 1, the model exhibits a constant or uniform influence. Thus, non-uniformity of influence is indicated by o ± 1.

In equation (7). a = o, NUI model reduces to NSRL model. Both the NUI and NSRL models can accommodate symmetric or nonsymmetric diffusion patterns and their point of inflection can occur at any point of time during the diffusion process. Also, these models can accommodate the effect of internal influence to be either increasing, decreasing or constant over time. Unfortunately, despite their advantages, the NUL and NSRL models are not amenable to simple explicit solutions.

Despite their increased flexibility these models are also characteristiczed by same seven assumptions underlying the fundamental diffusion model. Thus, further refined models were needed which could address to some of these assumptions. These types of models are (i) Dynamic diffusion models; (ii) Multi-innovation diffusion models; (iii) Space and time diffusion models; (iv) Multi-stage diffusion models; (v) Multi-adoption models; (iv) Multi-stage diffusion models; (v) Multi-adoption diffusion models; and (vi) Diffusion models with influencing or change agents.

Dynamic diffusion models :

In this category of models the unrealistic assumption of fixed $\bar{N}$ is relaxed. Mahajan and Peterson (1978) proposed a dynamic diffusion model where $\bar{N}$ is allowed to vary over time.

They specified

$\bar{N}(t) = f(\underline{s}(t))$

Where $\underline{S}(t)$ is a vector of potentially relevant exogenous and endogenous variables-controllable or uncontrollable-affecting $\bar{N}(t)$. Thus, we have a dynamic mixed-influence model of the form.

$$\frac{dN(t)}{dt} = \Big[a + bN(t)\Big]\Big[f(\underline{S(t)}) - N(t)\Big]$$

It is needless to say that the accuracy of the dynamic diffusion model largely depends upon identifying variables for $\underline{S}(t)$ that are closely related to or affect N(t). The nature of the innovation being investigated and the availability of the relevant data, and also the functional form of $f(\underline{S}(t))$ may vary across model applications,

In general, dynamic diffusion models give a satisfactory fit.

Multi-innovation models :

When an innovation is introduced in a social system, innovations existing in the social system may influence its diffusion. Peterson & Mahajan (1978) have identified four types of innovation interrelationship which could influence the adoption rate as well as the cumulative number of adoptions. These are :

(i) Independent-innovations are independent to each other but adoption of one may enhance adoption of others.

(ii) Complementary-increased adoption of one innovation result in increased adoptions of other innovations.

(iii) Contingent-adoption of one innovation is conditional on adoption of other innvations.

(iv) Substitute-increased adoptions of one innovation result in decreased adoptions of other innovations.

They extended the fundamental mixed-influence diffusion model to represent all four type of innovation interrelationships. The example of substitution relationship between two innovations '1' and '2' will illustrate the process :

For innovation 1 : $\frac{dN_1(t)}{dt}$

$$= (a_1 + b_1N_1 (t)-C_1 N_2 (t)) (N_1 (t)) \text{ .... } (9)$$

For innovation 2 :

$$\frac{dN_2(t)}{dt}$$

$$= (a_2 + b_2 N_2 (t)-C_2 N_1 (t)) (N_2-N_2 (t)) \text{ .... } (10)$$

Where C_1 and C_2 represent the hypothesized substitution effect of the innovations on each other.

In equation (9), the C_1N_2 (t) $[N_1-N_1 (t)]$ represents an interaction of adopters of innovation '1', resulting in a decrease in the rate of diffusion for innovation '1'.

Multi-innovation diffusion models can be used to test hypothesized relationship between innovations.

Space and time diffusion models :

Diffusion of any innovation simultaneously occurs in time and space. However, very few attempts have been made to integrate them.

One such attempt was made by Mahajan and Peterson (1979). They extended fundamental mixed-influence model to integrate the space and time dimensions, with the assumptions that (i) the growth in the number of adoptions in each region could be represented by mixed-influence model with its unique 'a', 'b' and N parameters and, (ii) the relative number of adoptions would be greater for those regions which are closer to the region where the innovation originated.

Mahajan and Peterson assumed that an innovation is first introduced into only one region, and the distance, x, between the innovative region and the remaining regions could be measure from the central point of one distribution to an other. This resulted in the model as.

$$\frac{dN(x,t)}{dt} = \left(a(X) + b(X)N(X\ t)\right)\left(\overline{N}(X) - N(X,t)\right)$$

Multi-stage diffusion models :

Earlier models considered the flow of social system members from potential adopters to current adopters-this may be considered as two-stage model. However, an adopting unit may pass through a series of stages in the innovation decision process. Let us, for example, consider three-stage diffusion models where uniformed social system members first become potential adopters and then current adopters. Let

X(t) = Number of social system members at time 't'.

Y(t) = Number of social system members who are aware of the innovation at time 't' but still have not adopted it.

Z(t) = Number of current adopters who have adopted the innovation.

It is assumed that the total population of the social system M remains constant over time, then at any time t,

$$X(t) + y(t) + x(t) = M$$

Or

$$\frac{dx(t)}{dt} + \frac{dy(t)}{dt} + \frac{dz(t)}{dt} = 0$$

$$\frac{dx(t)}{dt} = -\frac{dy(t)}{dt} + \frac{dz(t)}{dt} \quad \ldots\ldots\ldots\ldots\ldots\ldots\ldots\ldots\ldots \quad (12)$$

Thus, it requires the specification of two rate equations to describe the diffusion process.

The linkage between two-stage model and three-stage model can be drawn as :

$X(t + M \bar{N}(t);$

$Y(t) = \bar{N}(t) = N(t)$ and $Z(t) = \bar{N}(t)$

Multi adoption diffusion models :

This kind of model is relevant to product-innovation context where the successive increase in the number of adopters may consist of first-time buyers as well as repeat buyers of an innovation. As the repeat buyers tend to become intensive users of a product-innovation when they are satisfied with their experience at the initial adoption stage, it is the continued repurchase that creates a successful innovation. Thus, for non-repurchasable product-innovations the diffusion model depicts the first-purchase diffusion curve. For repurchasable products the purpose is to model the repeat purchase diffusion curve.

A number of repeat purchase models have been developed to forecast sales of product-innovations. Most of these models use survey data collected at test market stage to develop first-purchase and repeat purchase sales forecast.

Diffusion models with influencing change agents :

In this kind of models the effect of different strategies of innovation diffusion are incorporated. Most of the diffusion models contains coefficients of external influence a (t), and internal influence b(t), and the total number of potential adopters N(t) as parameters. Diffusion strategies may directly be incorporated into these models by representing these parameters as functions of relevant variables, for example :

$a(t) = A(\underline{S}(t)$

$b(t) = B(\underline{S}(t)$

$\bar{N}(t) = N(\underline{S}(t)$

Chow (1967) evaluated N as a function of price, Dodson and Muller (1978) represented N as a function of advertising. Mahajan and Peterson (1978) and Sharif and Ramanathan (1981) treated it as a function of social system population growth estimates. Lackman (1978) investigated the growth of new plastic in the automotive industry by representing $\bar{N}$ as a function of the ratio between corporate profits and corporate sales.

Economists and technologists took the coefficients of external influence and internal influence as functions of diffusion-influence and internal influence as functions of diffusion-influencing variables. They have studied by diffusionof technological innovations by taking the coefficient of internal influence as a function of profitability and investment. That is,

$b = Z - b_1 T - b_2 I$

where Z = a constant representative of an industry

T = a profitability index

I = an investment index

For most innovations, because of new technical advances and feedback from adopters, it is expected to change. Consequently, the impact of interpersonal communication on potential adopters may depend on the stage in the diffusion process in which it occurs.

Let us consider b N(t) $[\bar{N}-N(t)]$ in the internal-influence or mixed-influence models of diffusion.

We have $N(t) = \Sigma \left(N(j)-N(J01)\right) = \sum_{J}^{t} n(J)$

Where n(j) = No. of adopters adopting the innovation at time J.

Then $bN(t)\left(N-N(t)\right) = b(N-N(t)\sum_{J=1}^{t} n(j)$

Thus, the impact of interaction between n (1) $[\bar{N}-N(t)]$ and n(t)$[\bar{N}-N(t)]$, which is 'b', is assumed to be constant. Although some of the models considered earlier assumes the coefficient of internal influence, 'b', varies over time.

One approach of incorporating changes in technology and systematically varying the impact of interpersonal communication across the adopters, has been suggested byKalishand Lilien (1983). They proposed that the interaction term in the mixed influence model be expressed as,

$$\Sigma \text{ Wjn(j) } (\overline{N}\text{-N(t)})$$

$$\text{Where Wj} = \text{b.O(j)} \frac{1}{(1+r)^{t-j}}$$

Here O(j) = Perceived innovation quality at time j

r = intensity of influence of adopters at time j on potential adopters at time t.

= forgetting parameter (Kalish and Lilien 1983).

If r = 0 and 0(j) is constant, the model gives a constant impact interaction term.

The above review of the various types of diffusion models, though not comprehensive, clearly indicate that the available models cannot accommodate the relaxation of all the assumptions made regarding the models. Attention of the researchers is drawn in this area to develop more meaningful diffusion models.

References :

1. Bass, F. M. (1969), "A new product growth model for consumer durables." Management Sciences 15 : 215-227.
2. Bernhardt, I. and K. M. Mackenzie (1972), "Some problem in using diffusion models for new products". Management Science 19 : 187-200.
3. Blackman, A. W., Jr. (1974), "The market dynamic of technological substitution". Technological Forecasting and Social change 6 : 41-63.
4. Brown, L. A. (1981). Innovation Diffusion : A New Perspective. New York : Methuen.
5. Casetti, E. and R. K. Semple (1969), "Concernign and testing of spatial diffusion hypotheses", Geographical Analysis 1 : 254-259.
6. Chow, G. C. (1967), "Technological Change and the demand for computers", American Economic Review 57 : 1117-1130.
7. Dodson, D. A. and E. Muller (1978), "Models for new product diffusion through advertising and word-of-mouth" Management Science 24 : 1568-1578.
8. Easingwood, C. D., V. Mahajan and E. W. Muller (1981), "A nonsymmetric responding logistic model for forecasting technological substitution", Technological Forecasting and Social Change 20 : 199-213.

9. (1983) "A non-unform influence innovation diffusion model of new product acceptance", Marketing Science 2 : 273-296.
10. Griliches, Z. (1957), "Hybrid corn : an exploration in the economics of technological change", Econometrica 25 : 501-522.
11. Hagerstrand, T. (1967), Innovation Diffusion as a Spatial Process. Chicago : University of Chicago Press.
12. Kaliisha, S. and G. L. Lition (1983), A Market Entry Timing Model for New Technologies. Working paper, College Park : Pennsylvania state University.
13. Lackman, C. L. (1978), "Compartz Curve forecasting : a new product application", Journal of Market Research Society 20 : 45-57.
14. Mahajan, V. and R. A. Peterson (1978), "Innovation diffusion in a dynamic potential adopter population". Management Science 24 : 1589-1597.
15. (1979), "Integrating time and space in technological substitution models". Technological Forecasting and Social Change 14 : 127-145.
16. (1985), "Models for innovation diffusion". Sage University paper series Quantitative Application in Social Science, 07-048. Beverly Hills and London : Sage Pubns.
17. Mansfield, E. (1961), "Technical change and the rate of imitation". Econometric 29 : 741-766.
18. Paterson, R. A. and V. Mahajan (1978), "Multi-product models", pp. 201-231 in J. Sheth (ed.) Research in Marketing, Greenwhich, CT : JAI Press.
19. Rogers, E. M. (1983) Diffusion of Innovations, New York : Free Press.
20. Sahal, D. (1981), Patterns of Technological Innovation, Reading, M. A. : Addison Weeley.
21. Sharif, M. N. and C. Kabir (1976), "A generalized model for forecasting technological substitution". Technological Forecasting and Social Change 8 : 353-364.
22. Sharif, M. N. and K. Ramanathan (1981) "Binomial Innovation diffusion model with dynamic potential adopter population". Technological Forecasting and Social Change 20 : 63-87.

Chapter **34**

Path Model Explanaining Information Seeking Behaviour

D. K. Gogoi

The path analysis technique is a special type of multivariate analysis. The use of standardized multiple regression equation in examining a theoretical model is called path analysis (Loether and McTvish, 1976). It is a method that permits the social analyst to use explicit causal assumptions in the analysis of data (Mueller et.al. 1977). In India, the use of the path analysis technique has been very limited in communication researches and its sister disciplines, such an extension education, rural sociology, etc. Singh (1975) published a paper on path analysis technique in the India Journal of Extension Education. This was followed by the use of the technique by a number of researchers in agricultural communication and other areas of extension education (Chauhan and Singh, 1979; Sanoria and Singh, 1979, 1980, 1984; and Jhanisi Rani, et al., 1987). It appears from most of the published research articles cited above that the technique has been used by directly fitting it into the data without constructing a theoretical path model. In fact, the technique is applicable to test one or more theoretical models or causal structures of a set of variables which should be postulated prior to the empirical manipulation of data. Further, in these research studies, conclusions have been made regarding the population without testing the statistical significance of the path coefficients. There are available statistical test to examine the significance of a single or several path coefficients together.

In this paper, the technique has been explained in detail with the help of an example. A dependent variable, information seeking behaviour of farmers' has been considered as an example, to explain the technique. The description of the technique includes components of path analysis technique, assumptions, steps for formulating the causal model, estimating the path coefficients, "goodness of fit" of the path model and use of path analysis technique.

Components of path analysis technique :

There are five important components of the path analysis technique :

1. Causal model, path model or path diagram – The explicit statement of assumptions about the causal structure underlying a set of observed correlations is known as a causal model. The term path model is used when no feedback effects are postulated in a causal model. The causal model may be stated explicitly with the help of equations or path diagram. The path diagram is a graphic presentation of the causal model. It shows the direct effect and presumed in direct effects of the predictor variables on the dependent variables (s).

2. Predictor variables – The set of variables included in the model to explain the variable on the dependent variable are called predictor variables. The predictor variables are divided into two categories, exogenous and endogenous variables.

a. Exogeous variables – The variable which are not causally dependent on any other variables included in the model are called exogenous variables. They also are known as 'background' predictor variables. These must be temporarily prior to all other variables.

b. Endogenous variables – All other predictor variables (excluding the exogenous variables) are called endogenous variables. They are dependent on at least one other variable in the model.

3. Dependent variable – The variable to be explained in the model is called the dependent variable or the predicted variable. There may be more then one independent variable in a model.

4. Residual or error term – The effect of all other unspecified variables (not included in the model) are represented by error term. These are symbolized by 'e' or 'E' in the path diagram. In some literatures, the symbols 'R' (indicating residual term) or 'U' (indicating unique factor) also are used.

B. Independent/predictor variables :

In order to explain the information seeking behaviour 6 predictor variables were included viz. age, education, economic motivation, size of operational land holding, cosmopoliteness and media exposure. The measurement procedures of the independent variables are not reviewed here.

5. Path coefficient – It is the numerical value entered in the path diagram for each arrow connectingtwo variables. It is generally symbolized by Pij where 'j' is the predictor variable and 'I' is the dependent variable. The path coefficient is simply the regression coefficient using standardized varaibels.

Assumptions (Loether and McTavish, 1976)

A. Basic assumptions :

(1) Independent variables are related in a linear fashion with the dependent variable (s) and among themselves.

(2) Effects of independent variables can be added together to yield a prediction of the dependent variable (s).

(3) Independent variables are not correlated (there should be low intercorrelation).

(4) All variables are measure at the interval or ratio levels.

B. Added assumptions for testing a hypothesis about a population:

(5) The dependent variable (s) is/are normally distributed within categories of independent variables, singly or in combination.

(6) The variance in the dependent variable is equal across categories of independent variables.

C. Added assumptions for testing causal model :

(7) The theoretical ordering of independent and dependent variables should be known or at least can be hypothesized.

(8) The model should be a 'closed system'.

(9) Measures should have high (demonstrated) reliability and validity.

(10) Disturbance or error terms should be uncorrelated with each other or with independent variables directly connected to the same variable.

(11) There should not be a feedback effect or effects in the causal model. There should be unidirectional causality.

Step for formulating the causal model :

Path analysis focuses on the problem of interpretation and does not purport to be a method for discovering (Duncan, 1966). The use of the method requires a causal model to be tested.

A path model has been proposed to explain the dependent variable "information seeking behaviour of farmers". The model has been formulated following the steps given by O'Muircheartaigh and Payne (1976).

Selection and definition of the variables :

A. Dependent variable : "Information seeking behaviour". It is operationalized as the cumulative score obtained by a respondent in terms of his frequency of contact with specified sources of agricultural information, multiplied by the weightage given to the respective sources on the basis of their level of competency. A four-point scale was used to measure the frequency of contact with each source. The scoring procedure were as follws :

Response category	Score
Never	0
Once in a crop season	1
Once/twice in a month	2
Once in a week	3

The sources were assigned weightage on the basis of their level of competency as judged by few extension scientists. The ordering of sources were in conformity with the ranking of Murthy and Singh (1978). The weightages given to the sources were as follows :

(i)	Friends	1
(ii)	Progressive farmer	2
(iii)	VLW/Agricultural Sub-inspector personnel of private agencies	3
(iv)	Agricultural inspector/Agricultural officer (Block level)/Chief Agricultural Officer (district level)	4
(v)	Scientist of PAU	5

The information seeking behaviour score was calculated with the following formula

Information seeking

$$= \sum_{i=1}^{n} c_i \; w_i$$

Behaviour score

Where

C_i - Score for the frequency of contact with "i"th source of agricultural information.

W_i = Weightage given to "i"th source of agricultural information on the basis of level of competency.

n = Number of sources (in this case, n = 9).

Locate the exogenous and endogenous variables :

After critical examination of the set of predictor variables, age (X_1) and 'socio-economic status' (X_3) were identified as "exogenous". Theoretically, these two variables were found to be casually not dependent on any other variables included in the model. The other 4 variables viz., economic motivation, mass media exposure, cosmopoliteness and operational land holding remained as engogenous variables. The sequence of the predictor (engogenous) variables is in a clear temporal order. The order of the variable is -

(i)	Economic motivation	(X_3)
(ii)	Operational land holding	(X_4)
(iii)	Cosmopoliteness	(X_5)
(iv)	Mass-media exposure	(X_6)

Background hypothesized temporal order dependent variable of the endogenous variables

$$\begin{array}{l} X_1 \\ X_2 \quad X_3 \rightarrow X_4 \rightarrow X_5 \rightarrow X_6 \rightarrow (X_7)^* \end{array}$$

*in path analysis technique the dependent variable is marked as X_n i.e. part of a closed system of variables. Here, X_7 is the dependent variable.

4. Construct of the path diagram :

The diagram is constructed in a temporal sequence from left to right. The exogenous variables are connected by curved lines indicating simple association and direct lines with an one-headed arrow (signifying effects), are drawn from each predictor variable in the model to every subsequent variable. The error terms are shown by E's or e's connecting each with respective endogenous and dependent variable by one-headed arrows. The conventional nomenclature uses "Pij" to identify the path to variable X_i from the variable X_j within the context of the model and P_i to indicate the residual path from E_i to variable X_i.

To construct the causal model, particularly in ordering the endogenous variables (step 3) and proposing paths in the diagrams, we may use prior knowledge and/or common sense reasoning or may even hypothesize the direct and indirect effects through a number of variables. In the words of Li (1956) "........ the more we know the true relations among the variables the more meaninghful will be the results of path analysis". Figure 1 provides an example of a path model, showing the causal behaviour. The proposed path model is hypothesized causal structure of the predictor variables. One may modify or reformulate this model based on some theoretical basis and/or empirical evidences.

Estimating the path coefficients :

Singh and Choudhary (1979) gave three methods of estimating path coefficients :

(1) Use of regression equation :

Where b or j regression coefficient is the estimate of path coefficient.

(2) Use of simultaneous equations :

In this method, the basic theorem's equation is used : $r_{ij} = P_{ij} + P_{ik}r_{jk}$ read as $P_{ik}\, r_{jk}$ where I and j denote two variables in the model, k denotes the complete set of variables in the model with paths leading directly to the 'i'th variable.

(3) Matrix method/modified matrix method :

In this method, simultaneous equations are expressed in matrix form. The following are the equations expressed in matrix form for three independent variables on one dependent variable :

$$\begin{bmatrix} r14 \\ r24 \\ r34 \end{bmatrix} = \begin{bmatrix} r11 & r12 & r13 \\ r21 & r22 & r23 \\ r31 & r32 & r33 \end{bmatrix} \begin{bmatrix} P14 \\ P24 \\ P34 \end{bmatrix}$$

A = B X C .

C = B¹. A. Thus we require B¹ before estimating C ...

The path coefficient (Pij) calculated by one of the methods mentioned above are put on the respective arrow shanks of path diagram.

For estimating the path coefficients with the help of calculator, the matrix method is most convenient. In estimating the path coefficients of the present model, the following stepwise procedure was followed.

I. Preparation of inter correlation matrix.

II. Fit the equation (matrix method) to get the direct effect of the predictor variables. For this example –

r17		r11	r12	r13	r14	r15	r16		p71
r27		r21	r22	r23	r24	r25	r26		p72
r37	=	r31	32	r33	r34	r45	r36		p73
r47		r41	r42	r43	r44	r55	r46		p74
r57		r51	r52	r53	r54	r55	r56		p75
r67		r61	r62	r63	r64	r65	r66		p76

Solving the above equation, the direct effects of the predictor variables on the dependent variables were obtained.

III. Fit the equation considering the X_n - 1 variable i.e., X_6 as dependent variable. At this step, we exclude the variable, X_7. Then we get path coefficient of P_{61}, P_{62}, P_{63}, P_{64} and P_{65}. This procedure goes on until all the endogenous variables are exhausted. The path coefficients thus estimated are shown in figure 2.

Estimating the indirect effect :

To get the indirect effect of a variable, we have to multiply the path coefficient along the chain or path considered. For example, to see the indirect effect of Education (X2) through economic motivation (X_3) : $p_{32} \times p_{73}$ = .504 x .282 = .142. The indirect effects are usually given in a table.

"Goodness of Fit" of the path model :

Before examining the 'goodness of fit' of the path model, each of the path coefficients or several path coefficients together should be put to a 'test of significance'. O 'Muircheartaigh and Payne (1976) gave the following formula for testing the significance of the path coefficients :

(a) Testing one path coefficient

$$t = \frac{pij}{S.E.} \quad d.f. \quad (n\text{-}p\text{-}1)$$

Where pij = Path coefficient considered
S.E. = path coefficient's standard error
n = Number of cases (sample size)
p = Number of predictor variables

If the comptuted 't' is greater than that given in a two-tailed table of ' statistics for the desired level of significance and appropriate degrees of freedom, then it may be concluded that the path coefficient is significantly different from zero.

(b) Testing several coefficients :

$$F = \frac{(RSS_R - RSS_u)/k}{RSS_u/(n.p.1)}$$

Where RSS_R = Residual sum of squares of the restricted model
RSS_U = Residual sum of squares of unrestricted model
k = Number of restrictions (number of variables omitted)
p = Number of predictor variables
n = Number of observations
d.f. = Numerator = k; denominator = (n-p-I)

Land (1969) cited three general approaches to examine the 'goodness of fit; of one or more path modes :

1. One could examine the amount of variation in the dependent variable (s) explained by predictor variables as specified in the model.
2. One could examine the relative size and statistical significance of path coefficients to see whether they are large enough and significant to warrant inclusion of a variable or a particular path in the model may be examined by comparing the direct effects with indirect effects. Each path coefficient (pij) gives the direct effect of 'j'. The indirect effects can be obtained by multiplying the path coefficients along the causal chain considered.
3. One could evaluate the power of the model to predict correlation coefficients which are not used in computation of the path coefficients themselves. This criterion is ordinarily applied following exclusion of very small path coefficient in the fully recursive model by the second criterion (Mueller et.al. 1977).

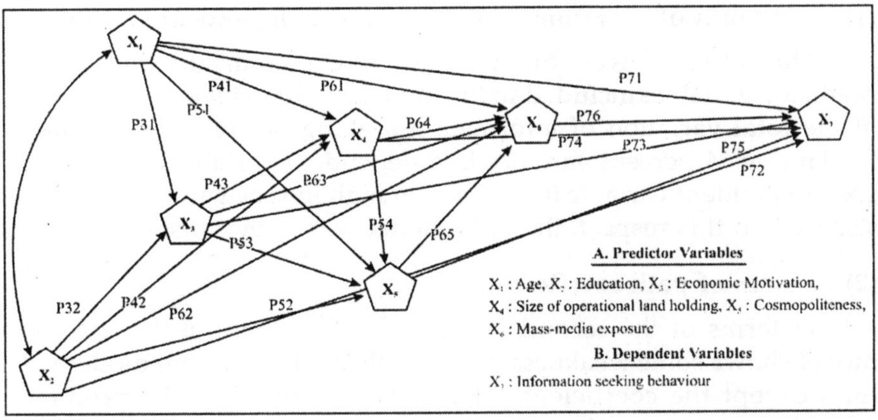

Fig. 1. Theoretical path model explaining information seeking behaviour

Discussion on the present path model :

In the present example, the first two criteria of "goodness of fit" were used to judge the appropriateness of the path model. The exclusion of one or more variables or paths from the model requires re-estimation of the path coefficients. The model has been examined by taking one criterion measure at a time.

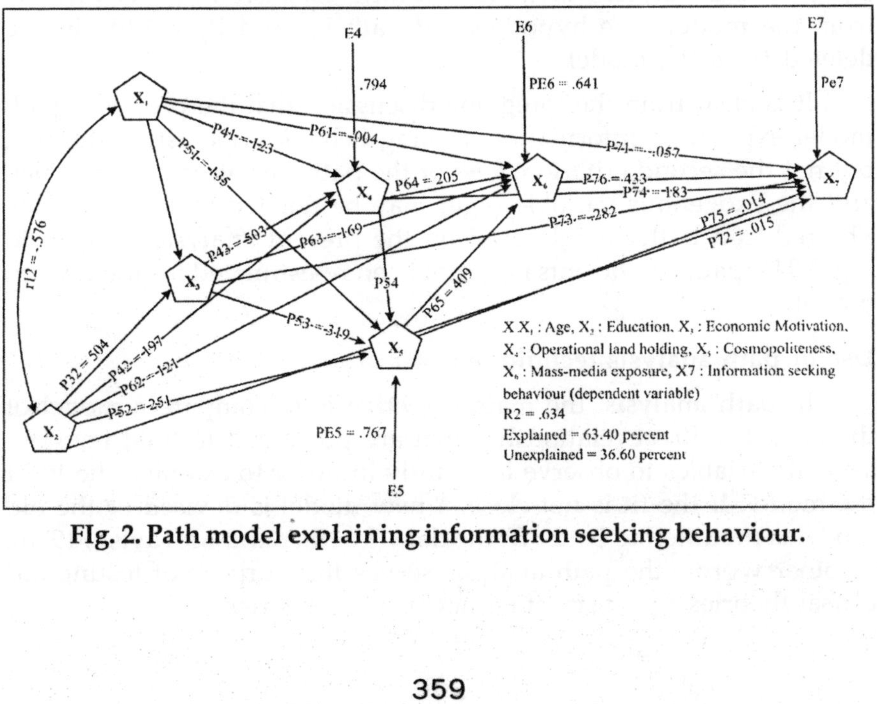

FIg. 2. Path model explaining information seeking behaviour.

(1) Amount of variation explained in the dependent variable :

The R2 value was found to be 0.634. This means that the six predictor variables included in the model, could explain 63.90 percent of the total variation of information seeking behaviour of farmers. Explaining 64 percent variance in a dependent variable with a set of six independent variable in reasonably high in social and behavioural sciences. In this respect, the path model was quite satisfactory.

(2) Sizes of path coefficients :

In terms of the second criterion of "Goodness of fit" the path model shows some weaknesses. The path coefficient were moderately high except the coefficients of P_{71}, P_{72}, P_{75} and P_{31}. The predictor variables viz;, economic motivation, size of operational land holding, and mass-media exposure were found to have significant, direct and positive effects on information seeking behaviour of farmers. The direct and indirect effects of the predictor variables on the dependent variable are given in Table-1. It is also from the table that two predictor variables, that is, education and cosmopoliteness have substantial indirect effects (>.100) on the dependent variable mediating through economic motivation and mass media exposure respectively. The size of correlation coefficient and also the direct and indirect effects of age on the dependent variable, warrants exclusion of this variable from the model. The hypothesized path P_{72} and P_{75} could also be deleted from the model.

It is clear from the foregoing discussion that the proposed path model explaining information seeking behaviour, satisfied the first but not the second criterion. Few of the paths, as already mentioned and the exogenous variable (age) may be deleted from the model. The reduced path model based on the present analysis is given in Fig. 3 The path coefficients have not been re-estimated for the revised model.

Use of path analysis technique :

In path analysis, the main objective is to compare a model of direct and indirect relationship that are presumed to hold between several variables to observe in a study in order to examine the fit of the model. If the fit is not close, a new model is devised or the old model is modified to better fit the data (Loether and McTavish, 1976). In other words, the path analysis serves the purpose of testing old causal theories or constructing new causal theories.

Table 1 : Correlation coefficients, direct and indirect effects of the predictor variables on information seeking behaviour

Predictor variable	Correlation coefficient	Direct effect	Substantial indirect effect : through Xi
X_1 : Age	-.209	-.057	-
X_2 : Education	.475**	.015	.142 : X_3
X_3 : Economic motivation	.662**	.252**	.100 : X_4
X_4 : Operational land holding	.592**	.182**	.100 : X_6
X_5 : Cosmopoliteness	.559**	.014	.177 : X_6
X_6 : Mass-media exposure	.720**	.433**	-

*Significant at .05 level

**Significant at .01 level

The path analysis technique has high information value than any of the statistical technique like zero order correlation or multiple regression analysis. The zero order correlations between a dependent and a set of independent variables permit researchers to order the variables according to the strength of their association. A multiple regression analysis gives us the effect of each independent variable and their cumulative effect on the dependent variable, but fails to give the causal structure of the predictor variables. It is here where the path analysis technique helps researchers by determining and testing the causal structure of the variables with estimates of direct and indirect effects.

References :

1. Ambasthaa, C. K. and Singh, K. N. (1980). "Intra personal communication pattern of extension system : a path analysis" Indian J. of Extn. Educ. XV, pp. 10-15.

2. Chauhan, K.N.K. and Sinha, B.P. (1979) "Effectiveness of television and its combinations in transferring technological know-how to farmers". Indian J. of Extn. Education XV, pp. 4-18.

3. Land Kenneth C. (1969). "Principles of path analysis" Chap. I in Edgar F. Borgatta (ed.). Sociological Methodology (1969). San Francisco, Jossey-bass.

4. Li, C. C. (1956) "The concept of path coefficient and its impact on population genetics". Biometrics, 12 pp. 190-210.

5. Loether, H. J. and McTavish, D.G. (1977). Descriptive statistics for sociologists. Boston, Allyn and Bacon Inc. pp. 321-328.

6. Muller, J. H. et.al. (1977). Statistical reasoning in sociology. Boston, Hoffton Mifflin Co. pp. 312-330.
7. Murthy, A. S. and S. N. Singh (1974). Communication behaviour of farmers, Delhi : New Heights.
8. O' Muirchearcheartaigh, C. A. and payne, C. (1976). Model fitting (vol. 2). New York, John Willy & Sons, pp. 60-64, 81-104.
9. Sanoria, Y. C. and Singh, K. N. (1979). "Communication pattern of extension personnel - a path analysis". Indian J. of Extn. Education, XV, pp. 19-25.
10. (1980). "Communication pattern of agricultural scientist : A path analysis. "Indian J. of Extn. : Education XVI, pp. 39-45.
11. Jhansi Rani, G. et.al. (1987) "Influence of selected variable on scientific productivity of agricultural scientists". Idnian J. of Extn. Education, XXIII, pp. 26-29.
12. Singh, B. K. (1975). "Path analysis is social research". Indian J. of Extn. Education XI : pp. 54-62.
13. Singh, R. K. and Choudhary B. B. (1979). Biometrical methods in quantitative genetic analysis. Ludhiana : Kalyani publishers. Pp. 70-79.3
14. Tyagi, K. C. and Sohal, T. S. (1981) : "Antecedents of communication dairy innovations". Indian J. of Extn. Education, XVII : pp.1.
15. (1983), "Antecedents of dairy farmers' empathy". Indian J. of Extn. Education. XIX : pp. 51-56.
16. (1984) "Factors associated with adoption of dairy innovation", Indian J. of Extn. Education, XX, pp. 1-8.

Chapter 35

Information and Communication Technology (ICT) : An Emerging Tool for Rural Development

Dhanraj A. Patil, Jayawant B. Ambekar

The village epitomize the sole of India. With more than 70% of the Indian population living in rural areas, rural India reflects the very essence of Indian culture and tradition. A holistic development of India as a nation rests on a sustained and holistic development of rural India. Information and communication technology (ICT) has emerged as an effective facilitator in the development of any society and is a prime drivingforce in the growth of economies worldwide. In this context, ICT have lot to offer to develop rural sector. This paper provides an account of some of the major initiatives particularly touching to the rural domain. It also highlights the impact of ICT efforts on the rural society.

The most conspicuous development in the globalization context has been the information and communication technology(ICT) influx. It seems to have turned the world into a global village. Hence, the developing world is gradually catching up with the technological advancement to solve its socio-economic problems. Even though the benefits of ICT come with the tag of a number of constraints, it is being considered as a panacea for all ills. In a vibrant democracy like India which is marked by plethora of languages and diverse cultures ICT assumes greater significance to develop the rural sector. Present paper highlights the impact of ICT efforts on agriculture, health, women empowerment and panchayat raj etc. It also explores the hurdles involved in rural ICT projects, finally it gives some vital suggestions in this regards.

Methodology :

(a) Key concepts :

(1) ICT (Information and Communication Technology) :

The organization for Economic Co-operation and Development (OECD) defines "ICT as the set of activities that facilitates, by electronic means, the capturing, storage, procesings, transmission and display of information."

(2) Rural Development :

In this study rural development means, to promote the participation of rural people in ICT projects and programmes designed to assist their development and develop their Information, Education and Communication base of rural people, thereby create awareness among rural people of their problems and to build up their solidarity to tackle key problems in agriculture, health, education, women empowerment, panchayat raj, co-operative sector etc.

(b) Objectives :

The specific objectives of the present paper are :

(i) To elucidate the role of some of the major ICT initatives in rural development.

(ii) To study the impact of ICT efforts on agriculture, health, women empowerment and panchayati raj.

(iii) To print out the impediments for the effective implementation of ICT based project.

(c) Data :

The present paper is based mainly on secondary data. However, it also based partially on the primary data of the Warana Wired Village Project, Warananagar, District Kolhapur.

Historical roots of ICT in India :

In India, the path towards technology-induced development, especially associated with ICT, was given a vent in 1984 by the Congress Government under Rajiv Gandhi and the technologist Sam Pitroda. They assumed power and adopted informatization of Indian society as an effective route to development (See, Singhal and Rogers, 2001). A high power National Task Force on Information Technology and Software Development was later set up in 1998 by the BJP

Government. This Task Force prepared the blue print for making the adoption of IT as a National movement, by establishing a wide network of empowered taskforce at all governmental and non-governmental levels. The New Telecommunication Policy 1999 has stipulated targets in terms of establishing a telecom network with a view to achieving tele-density of 7 percent by 2005 and 15 percent by 2010 for a knowledge society (Bajwa, 2006 : 49).

The concept of knowledge society means that it will incorporate new knowledge will takeover the role of capital and labour in the economy, further leading to structural and functional changes in family health and education. Another vital questionis how can these various structural change facilite the version institutions and their adoption?

Table-1. Some innovative rural ICT-based initiative in India

Initiative	Sponsoring agency	Important features	Major thrust or area or level
Public sector			
Cyber extension project	National Institute of Agril. Extension Management (MANAGE), Hyderabad	Village information centers, institutional support to other ICT projects	Post harvest processing women and child welfare
Agriculture techn. Information centre (ATICs)	Indian Council of Agriculture Research (ICAR), New Delhi	Call centre, expert systems, researchers, farmers linkage	Agronomic practices, technologies inputs
Agricultural gateway of India	National Academy of Agricultural Research Management (NAARM), (ICAR), Hyderabad	Helpline service through telephone, researcher-farmers linkage	Cultivation practices, plant protection, new technologies
Helpline service	Chandra Shekhar Azad University of Agriculture and Technology (CSAUAT), Kanpur	Helpline service through telephone, researcher-farmers linkage	Cultivation practices, plant protection new technologies
Warna wired village pojrect	NIC, Warna Cooperative Complex, Warna, Dist., Kolhapur (Maharashtra)	Wide Area Network Computer booths	Cultivation practices, pest and diseases, marketing, processing etc.

Initiative	Sponsoring agency	Important features	Major thrust or area or level
Gyandoot project	Gram panchayat, Dhar district, Mahdya Pradesh	Soochanalayas, Portal Partial recovery, Panchayat community partenership	Agriculture produce, market intelligence, auction rates, land records
Interactive touch screen information kiosk	Dept. of International Development (DFID) Rajiv Gandhi College of Animal Sciences (Pondicherry), University of Redading (U.K.)	Information Kiosks text, pictures with sound backup	Information about important cattle diseases and mgt. of cattle
Lokmitra project	National Informatics Centre, Gram Panchayat, Hamirpur district, H.P.	Lokmitra Soochana Kendra, in 25 Panchayats	Classified components, job vacancies, market rates, tenders etc.
E-choupal	Indian Tobacco Company	e-choupal, crop specific internet	Market price, cultivation, practices, wheather etc.
Rural information kiosks	EID Parry, Tamil Nadu	Farmers managed kiosks portal, public private partnership in applying ICT improvattions and content development	Farm practices, farm advisory, maker intelligence, weather etc.
I-kisan portal	Nagarjuna group, A.P.	Portal in regional language	Agricultural practices, plant protection, animal husbandry, weather etc.
NGOs			
Village Knowledge Centre	MS Swamination Research Foundation (MSSRF), Chennai	Knowledge centre, pro-poor, pro-nature pro-women and community ownership	Agronomic practices cattle feed and weather, schemes
Computer on wheels	Pingali Rajeshwari entrepreneur; A.P.	Internet on moto bikes, portals, remote area, resource poor	Market, weather, plant protection and animal health
www.doctoranywhere.com	Tata Council for Community Initiatives (TCCI) Pune (MS)	Telemedicine Service through websites and computer	Medicinal council, medicinal treatment in PHC

Source : Abhigaru et. al. (2003), Patil, D.A. (2006)

Table-2. Reduction in transaction cost for accessing information

Particulars	e-choupal			I-kisan			Helpline		
	Without ICT	With ICT	Reduction (%)	Without ICT	With ICT	Reduction (%)	Without ICT	With ICT	Reduction (%)
Labour cost per visit* (A). Rs.	34.12	6.18	81.89	41.56	3.28	92.10	35	2.59	92.60
Travel cost per visit (B) Rs.	13.87	0	100	9.25	0	100	5.88	0	92.58
Total expenditure per visit (A+B) Rs.	47.99	6.18	87.12	50.81	3.28	93.54	40.88	2.59	93.66

Source : Abhiguru et. al. (2003)

*Note : Imputed value of labour hours.

Table-3. How e-choupal saves costs (Soyabean trading)

	Nature	Mandi	e-choupal
Cost incurred by farmers (Rs./tonne)	Trolly freight	120	120
	Labour	50	Nill
Cost incurred by processor (Rs./tonne)	Middlemen	150	Nill
	Handling loss	50	Nill
	Total	370	120
	Commission	100	50
	Gunny bags	75	75
	Freight to factory	120	Nill
	Storage and handling	40	40
	Disbursement cost	Nill	50
	Total	335	215
	Total chain	705	335

Source : Saran, 2004

Table-4. Time and money saved through village knowledge centres

Group/category	Number	Distance saved (kms.)	Time saved (hours)	Money saved (Rs.)
Total number of users	2518	74304	2477	12384
Men	2064	61560	2052	10260
Women	354	12744	45	2124
Below poverty line	85	3060	66	66
Illiterates	20	720	24	120
First times users	625	22500	452	3450
Men	425	15300	510	2550
Women	200	4200	340	1200

Source : UNESCO, 2001

Table-5. Constraints experienced by the beneficiaries

Constraints	Organization					
	E-C	IK	HL	COW	GD	MAN
Insufficient regional specific information	25	55	-	40	-	50
Inadequate subject matter	25	10	37.5	62.5	-	45
Not suitable to all kinds of information	20	-	30	-	-	-
Inadequate support from facilitator	10	35	-	-	-	37.5
Inadequate facilitator's knowledge	52.5	30	-	25	-	35
Facilitator is not available	-	-	50	-	-	-
Lack of infrastructure facility	-	62.5	-	-	55	30
Inadequate internet phone connectivity	-	50	50	-	50	47.5

E-C : E-choupal, IK : I-kisan, HL : Helpline service, COW : Computer on wheels, G.D. : Gyandoot, MAN : MANAGE

Source : Adhiguru et. al. 2004

Table-6. Hurdles involved in W.W.V.C.P.

S. No.	Hurdles in W.W.V.C.P.	Experimental village		Rank
		Frequency	Percentage	
1.	Insufficient specific information	**32**	**32.00**	I
2.	Irregular availability of operator	**16**	**16.00**	III
3.	Lack of infrastructure facility	**11**	**11.00**	V
4.	Inadequate internet/phone connectivity	**21**	**21.00**	II
5.	Irregular electricity	**13**	**13.00**	IV
6.	Inadequate operator's knowledge	**7**	**7.00**	VI
Total		**100**	**100.00**	

Source : Patil D. A. (2006)

ICT : Some prominent initiatives :

Let us now compare the ICT initiative in India in terms of Public vs Private/NGOs, their sponsors, features and their major thrust of development. (Some examples of such initiatives are listed in Table 1).

(1) ICT for agriculture :

(i) **e-choupal** : Indian Tobacco Company (ITC's) International business division is one of India's largest exporters of agricultural commodities has conceived e-choupal launched in June 2000 as a more efficient supply chain aimed at delivering value to its customers around the world on a sustainable basis. ITC's e-choupal is a unique example of using ITC's for agricultural development, e-choupal has already become the largest initiative among all internet based interventions in rural India, e-choupal services today reach out to more than 3.5 million farmers growing a range of crops - Soyabean, Coffee, Wheat, Rice, Pulses, Shrimp - in over 31,000 villages through 5200 Information Kisoks across six states (Madhya Pradesh, Karnataka, Andhra Pradesh, Uttar Pradesh, Maharashtra and Rajasthan).

(ii) **I-Kisan Portal (Andhra Pradesh)** : TheNagarjuna Fertilizers Company Limited (NFCL) is a agribusiness based at Andhra Pradesh. They are disseminating various farming information to the farmers at various places through ICT centers. To do this job, they have engaged 200 filed assistants and 25 technical assistant to serve as contact interface between company and the farmers. They also have Krishi Vidnaynan Kendra (KVK) at Kakinada (Andhra Pradesh).

(2) ICT : For livestock :

India has a huge bovine population of 196 million cattle and 80 million buffaloes together producing 84 million tons of milk accounting for about 13 percent of the world's milk production. The challenge for veterinary institutions to reach millions of livestock owners spread in length and breath of the country remained insurmountable. Thus, ICT initiative is expected to provide a strong case for effective coverage and dissemination of livestock out reach information to the farming community. In this context, three e-mail conferencing system, viz., goat-net-cirg.nic.in, livestock-net@cirg.nic.in and fishnet@cirg.nic.in, have been launched by the institute to help information inflow among technologies, farmers, development officers and planners (Singhand Radhika, 2003 : 2). Under the Animal Health Project funded by Department for International Development (DFID), Rajiv Gandhi College of Veterinary and Animal Science Pondicherry in collaboration with University of Reading, UK, has installed a Interactive Touch Screen Information Kiosk in one of the rural regions of Pondicherry on a trial basis. It has information on important cattle diseases in addition to management of cattle and methods of acquiring information. Illiterate livestock keepers can easily access the information with touch of the screen, which has text and pictures with sound backup.

(3) ICT : For Health :

(i) Telemedicine service in Pune Primary Health Centers :

Those living in interior Pune village will now be able to avail expert medical council well within their means. The Pune district administration has teamed up with www.doctoranywhere.com and Tata Council for Community Initiatives (TCCI) to launch a telemedicine service from a government Primary Healtcare Centre (PHC). The service, says the Chief Executive Officer of Pune district administration V. Radha, will reduce the traveling time and expenditure of the poor villagers. She further says, "if we have the headquarters connected with these PHC's, we can respond immediately. If there is an emergency, we can at least rush for medicines. There are at least five to six doctors always present at the headquareters who can respond even if there are no specialists (www.doctoranywhere.com). One another telemedicine project was started in 2000 by Apollo Hospitals. The project has already benefited over 3749 patients at 11 remote locations in India. The Indian Space Research Organization (ISRO) and Development and Educational Communication Unit (EDCU) has initiated several telemedicine projects in remote areas. Sankara Nethralaya is asuper-speciality eye hospital (Chennai) is one of the hospital chosen by ISRO for the pilot project providing telemedicine though satellite connectivity. The mobile tele-ophthalmology unit also provides continuing medical education programmes for doctors and create awareness about eye-diseases among rural people 9Arul, 2005 : 110).

(4) ICT : For panchayat raj :

(i) Lokmitra project launchedin May 2001 in the Hamirpur district of Himachal Pradesh Lokmitra Soochana Kendras were set up at 25 centrally located panchayats. National Informatic Center (NIC) provided the Lokmitra Software and necessary training to the Soochaks (operators). The services includes classified complaints, forms download, job vacancies, tenders, schemes information, market rates, etc. Some of the other initiative's are GAYNDOOT in Dhar District of Madhya Pradesh and SAYTU in Maharashtra are successfully implemented.

(ii) **E-Gram Suvida :** E-Gram Suvida is Geographical Information System (GIS) based on Management Information System that links the maps of village resolution with the associated data to generate thematic maps based on users queries. An easy to use software, e-gram Suvidha is currently operational in five districts of Madhya Pradesh (www.nic.in).

(5) ICT : for women's empowerment :

Women being a vital part of Indian economy constitutes about one third of national labour force. More than 90 percent of rural women in India are unskilled (Menambiguti et. al. 2004 : 2). Proper training to the women and imparting skills through the use of ICT will be quite helpful. In this direction, M. S. Swaminathan's Pondicherry based project is a unique example. The Information Village Project has been implemented by the M. S. Swaminathan Research Foundation (MSSRF) in collaboration with the International Development Research Centre (IDRC). The specific objectives of this project is to train youth especially women, in rural areas in operating information shops. The info villages provide local-specific information to villagers related to prices of agricultural input (seeds, fertilizers, pesticides), agricultural output (rice, vegetables) market, healthcare, cattle disease, transport, weather, etc. in local i.e. Tamil language. MSSRF provides this service free of cost and it has won two international awards.

(6) ICT : For co-operative development :

(i) **Dairy co-operative :** National Dairy Development Board, ICT is being used at milk collection centers and in co-operatives to measure butter fat content milk, test the quality of the milk and helps to make payment to the farmers. As a result, it reduced time for payments and instilled confidence among the farmers in the co-operative set up. Amul Co-operative has introduced a number of ICT based operations, which have contributed to efficiency of milk procurement and distribution. The NDDB has also been able to deal with problems of unethical practices in the milk enterprises (Bhatnagar and Schware, 2005 : 51).

(ii) Sugar Co-operatives: The Warna Wired Village Project, covering 70 villages in Kolhapur and Sanglli district of Maharashtra state. The village information center has providing indepth information about sugarcane cultivation and marketing to the farmers (Patil, 2006 : 3).

(iii) Impact of ICT initiatives :

In the above sections, a few cases have been discussed. The impact of some of these initiative is presented below based on the experiences of villages and farmers, as reported by (Adhiguru et. al. 2003 : 117). Segregating the exclusive impact of ICT is difficult but it broadly indicates how ICT can be one of the significant inputs in the development paradigm.

(1) Reduction in transaction cost :

Generally, for receiving information or any technical knowledge on crop cultivation, farmers go to a nearby taluka headquarters, where they can meet officials in the Department of Agriculture or get advice on purchasing agro-inputs from private dealers. Usually, private dealers do not spend time in offering technical advice to a farmer unless he/she actually purchase a product. In other words, it is product-oriented why farmers have failed to increase the overall efficiency in production. The ICT-enabled initiative could change the scenario and the farmers could get technical advice by visiting kiosks located in their village or nearby village and save transaction costs as well. The transaction cost incurred by the farmers was worked by computing variables like distance traveled, labour hours involved per visit, cost equivalent of labour hours, and traveling cost per visit. It is very clear from Table no. 2 that through ICT enabled services, a farmer could save this traveling time and transaction costs by more than 80-90 percent.

For marketing the produce as well, the farmers used to spend considerable time and money for price discovery and later for selling the produce. Most of the time, they got a price disproportionate to the quality of the farm produce. The e-choupal (soyachoupal) concept is a live example of how ICT can help overcome these difficulties. By using soyachoupal, the farmers could save 68 percent of the transaction costs due to information-led decision making in when and where to sell their farm Soyabean produce while the processing company could save about 36 percent on account of saving charges towards freight to processing factory (see Table no. 3).

The farmers also reported that they had registered an increase in the sale value for their soyabean produce by around two percent due to the premium price offered by soyachoupal for their qulity farm produce (ibit : 17) Farmers, especially smallholders could minimize the difficulties and the cost involved in the process of price discovery and the purchase of farm inputs by availing another distinct service offered by soyachoupal. Farmers could place their requirement of inputs to the Sanchalak, who place bulk demand with the ITC. ITC arranged to deliver the products, ensuring quality farm inputs at reduced transaction costs.

One-fourth of the respondents contacted under the study reported that they could reduce the cost of procurement of inputs by 4 percent. Though Gyandoot kiosks, farmers could get the land records in village itself and save transaction costs by 66 percent. The could save time, travel charges and avoid unecessary administrative hurdles (ibid : 20). Through Gyandoot, the farmers, particularly smallholders, have been getting higher returns on the basis of the timely access to market information. In one specific case, the smallholders could get Rs. 4000 per ton in a nearby market for the Potatoes they were earlier selling at Rs. 3000 per ton though the local middlemen (Economic Times, 2002).

(2) Time saving :

According to a study by UNESCO in 2001, the users in the three knowledge centers of MSSRF, Pondicherry have saved a travel distance of around 100,000 kms, and 4000 hours amounting to imputed value of Rs. 16,000 (see Table No. 4). Time is a valuable commodity for the landless labourers and women in the rural areas (particularly during the working seasons) and therefore, time saved means money-earned. Further, it also leads to qualitative social development in the health and education.

Awareness and utilization of entitlements :

Through the village knowledge centre, Panchali, a farm-woman living in Kizhoor village, of Pondicherry came to know about the availability of subsidized seeds at PASIC (Pondicherry Agro-Service and Industries Corporation Limited) and procured 40 kg of white ponny Paddy variety seeds at the rate of Rs. 10.50 per kg. This timely availability of information saved her both time and money (Rs. 80, subsidy for 40 kg seeds) (UNESCO, 2001). Thus, it should be noted that the assetless, ultra-poor families are among the major users of the village knowledge centre. Access to and control of information by poor women is a step towards empowerment.

Impediments in the use of ICT :

In India, the utilization of ICT in agriculture and rural development is in the take off stage and obviously there are many field problems (see Table No. 5). For instance, among the various constraints reported by the beneficiaries in different initiatives subject inadequately and lack of content in local language were found to be prominent constraints. The facilitator's co-operation and his/her subject matter knowledge was perceived to be a major factor as well. About half of the respondents expressed irregular internet connectivity as one of the major constraints, except in the case of initiatives like e-choupal and Computer on Wheels (COW). These two initiatives use solar cell and battery back up for the regular power supply and the e-choupal has VSAT connectivity whereas COW and GPRS (General Package Radios System) connectivity. Farmers faced problems in using the helplines due to very low teledensity in rural Uttar Pradesh (0.56 phones per 100 people by March, 2003).

However, Patil D. A. (2006) in his study on Warana Wired Village Communication Project (WWVCP) in Kolhapur district of Maharashtra, also found similar findings in this regard (see Table No. 6).

It is widely excepted that ICT's are playing an important role in the development process. But at the same time the most difficult stage for any ICT initiative, is to sustain them. In the light of the above discussion, Table No. 6 shows some of the key hurdles involved in WWVP for instance, among the various hurdles reported by the peasants of experimental village, insufficient specific information (32 percent) and inadequate internet/phone connectivity (21 percent) were found to be prominent hurdles. The irregular availability of operator and electricity was perceived to be a major factor as well. About 18 percent of the peasants opined that lack of infrastructure facility and inadequate operators knowledge as one of the major hurdles involved in WWVP.

He also observed that, in Warna Wired Village Project (Maharashtra) the information kiosk are mostly accessed by members of the co-operative farmers who own their land. But the poorest, landless labourer currently do not have reason to visit the kiosks.

Suggestions for effective utilization of ICT :

Several studies (Adhiguru et. al. : 2004, Hiremath, et. al. : 1996, Cechinis and Raina : 2002, Patil : 2006, Dhemeja and Mendury : 2002 and Bejwa : 2005) suggested following strategies for effective utilization of ICT projects especially in rural areas.

(1) **Content development :** Generic and indigenous knowledge, success stories in local language to meet local needs.

(2) **Role of universities and institutions :** Research and development institutes, Universities must conduct ICT based project and also conduct Ph.D. and M.Phil. research on this subject.

(3) **Integrated approach :** There is need for an integrated approach, based on the real information needs of the potential users, and not on mere 'technical fix'.

(4) **Enabling environment for IT :** There is need to create an information friendly environment, characterized by telecommunications reform and information policies, the track of creating such environment falls not to one but all the partners for development.

(5) **Holistic perspective :** We need to place ICT in a holistic perspective of Information System, as social systems are rooted in a context of people and social structure.

(6) **Development of software :** To develop capacity of managers/ operators of information services to identify and supply the information needs of users and human capacity (literacy, numeracy) of beneficiaries to move use of information.

(7) **Bridging the divide :** Strategic development of ICT to address socio-economic dividers : age, gender, community, land holding and geography is eseentail.

(8) **Low-cost training institutes :** School and Panchayat Offices should impart ICT training. Students can be effective 'no cost trainers' NSS students can play key role in this context.

(9) **Learning from success stories :** Success stories provide a forum for discussing and evaluating international experience that contribute to lesson-learning constitute best-practices.

Concluding remarks :

The use of ICT in agriculture in general and rural development in particular remains restricted in India. Effective utilization of ICT has the potential to make the rural communities in India prosperous. All efforts should be to guide the ICT initiatives with a focus on the socially deprived and economically disadvantaged sections that have not received adequate attention so far. Information systems are made of people and social structures and hence should be rooted in them. The narrow ICT coverage is found to be financially non-viable. Further, no single agency can effectively deliver this critical input. Besides the Public sector, the need for a proactive participation by the private sector, NGOs and other civil society organizations is being increasingly felt. For ICT initiatives to be successful and sustainable in the long run, collaborative efforts are indispensable. ICT can definitely play key role in this endeavour if the local Panchayats are encouraged and supported to set up ICT based village development information centres.

Note :

1. India is a developing nation that is making remarkable progress toward development through the informatization strategy. Informatization is the process through which communication technologies are used as means of furthering socio-economic development technologies are used as a means of furthering socio-economic development (Singhal and Rogers, 2001 : 18).
2. The aims of the task force was to formulate the draft of a national policy on Informatics to enable India to emerge as in Information Technology super Power within the next ten years.
3. Task force June (2001), Steering Committee Members consisted of Dr. A.P.J. Kalam, Dr. R. A. Mashelkar, M. K. Kaw, Dr. K. Venkatasubraminam, Prof. A. Parthasarthi, 14 members and 3 co-opted members.
4. KIOSKS is Internet connectivity for small towns and villages, that is just similar to the STD booths which revolutionized communication by connecting villages to the outside world, through technological devices such as computer, VSAT (Very Small Aperture Terminals), WLL (Wireless Local Loop), LAN (Local Area Network), WAN (Wide Area Network).
5. The farmer-owned Amul Co-operative in Anand has shown that since the 1960s an integrated approach along co-operative lines could enhance production and marketing of milk. It has

become a model for most dairy development projects in India.

References :

1. Adhiguru et. al. (2003) Project on "Innovative Institute for Agricultural Technology Dissemination : Role of ICT", National Center for Agricultural Economics and Policy Research, New Delhi, pp. 117.
2. Abhiguru, P. and Mruthyunjaya (2004). "Institutional Innovations for using Information and Communication Technology". Policy Brief 18, National Centre for Agricultural Economics and Policy Research, New Delhi, pp. 17-22.
3. Arul, A. (2005). "Satellite Based Telemedicine Connectivity for Eye Care in India," Media Asia, Vol. 32, Singapore. Pp. 110.
4. Bajwa, S. B. (2006). "ICT Policy in India in the ear of Liberalization : Its Impact and Consequences." GBER, Vol. 3 No. 2 pp. 49.
5. Bhatnagar, S. and Schware, R. (2000). Information and Communication Technology in Development. New Delhi : Sage Publications. Pp.51.
6. Cecchini, S. (2002). "Information and Communication Technology for Poverty Reduction in Rural India." World Bank : Washington D. C.
7. Dhameja, A. and Medury, V. (2002) "Information and Communication Technology in the Globalization Era : The Socio-economics Concerns," New Delhi : Indira Gandhi National Open University.
8. Hiraemath, B. N. and V. Ballabh (1996). "Integrating Small Farmers with Market Problems and Prospects in Small Farm Diversification." Ed. T. Haque, New Delhi : National Centre for Agriculture Economics and Policy Research.
9. Meebambigati, G. et. al. (2004). "Role of Communication Technonology in Rural Development," Journal of Communication Studies. Vol. 23. Varanasi (U.P.). pp.2.
10. Patil D. A. (2006). "New Communication Perspective for Cooperative and Rural Development," Journal of Communication Studies. Vol. 25, Varanasi (U.P.). pp. 3.
11. Patil D. A. (2006). A Study of Communication Networks and Development in Kolhapur District. Ph.D. thesis (Sociology), submitted to Shivaji University, Kolhapur.
12. Saran, R. (2004). "To Boldly Go Where." India Today, December, 13, pp. 12.
13. Singh S. K. and Radhika V. (2002). "Development of e-mail conferencing system for improvement of livestock through ICT." Paper presented in International Congress, Asian-Australian Association of Animal Production Societies, 23-27 September, 2002 at New Delhi.

Web-site :

1. www.manage.gov.in
2. www.doctoranywhere.com
3. www.nic.in
4. Economics Times (2002) http://news.com/India's+Software.html.
5. UNESCO (2001) www.unesco.or.id/apgest/india-bp-it-pdt.

Chapter **36**

Traditional Folk Media and Its Importance in Development Communication

O.P. Mishra, Pooja Sah

Folk media or traditional media often appeared as one of the viable ways of solving present problems in development. Local or traditional communication channels such as puppet show, folk drama, story telling, community teahouse, open market, and village meetings are available to all and enjoyed by person of different age groups, generally at a very low cost. The Government of India made deliberate efforts to use the traditional folk media for disseminating important themes like national integration, communal harmony, health and family welfare, girl child, prevention of drug abuse etc. The paper discusses about concept, types, advantages and importance of folk media in disseminating development messages to rural people.

Indigenous communication systems may be defined as "a communication system embedded in the culture which existed before the arrival of mass media, and still exists in many parts of the world with a certain degree of continuity, despite changes, as a vital most of communication" (Wang and Dissanayake, 1982). It includes local or traditional communication channels such as puppet show, folk drama, story telling, community teahouse, open market, and village meetings. Traditional communication media like folk songs, drama etc. are highly structured and deeply rooted in the socio-cultural mores of the society. Proverbs, riddles, folktales, myths and folk songs have played an important role in the development of society. Village festivals can serve as occasions for political education, specific development issues can be scripted into folk debates and desirable value system can be woven into folk drama (Patron, 1981). Indigenous communication channels are available in many forms such as folk media, indigenous organization and social gatherings, deliberate instruction, record and unstructured channels (Mundy and Compton, 1991).

Folk media may constitute a part of the indigenous communication system but there are also other communication media and channels which serve similar function and are frequently discussed in the same context like meeting places such as festival; informal organization and institution such as water sharing system and mother clubs (Wang and Dissanayake, 1982).

Types of traditional media :

The traditional media can be classified into four major forms :

1. Folk theatre forms :

- Tamasha (Maharashtra)
- Powada or pawada (Maharashtra)
- Dangi tamasha (Gujarat)
- Dashavatar (South Konkan, Maharashtra)
- Nautanki (North India)
- Yakshagana (Karnataka)
- Jatra (Bengal and Orissa)
- Bhavai (Gujarat)
- Therukoothu (Tamilnadu)
- Burra Katha (Andhra Pradesh)
- Ram Leela (Uttar Pradesh)
- Bidesia (Bihar)
- JatJatin (Bihar)

2. Keertana (Harikatha or Harikeertan)

i. Keertana (Maharashtra)

- The traditional or Naraditya Harikatha
- The Warkari Harikatha
- The Nationalist Harikatha

ii. Keertana (Gujarat)

3. Folk songs/folk verse/ballads/folk saying and riddles

There are more than 300 folk musical styles in India of which the more prominent are :

- The Baul and Bhaitali of West Bengal,
- The Doha and Garba of Gujarat,

- The Chaiti and Kaijri of Uttar Pradesh
- The Kolkali pattus of Kerala,
- The Bihu of Assam,
- The Mnad and Panihari of Rajasthan,
- The Rauf and Chhakri of Kashmir,
- The Sua and Dadaria of Madhya Pradesh,
- The Mando and Dhalo of Goa,
- The Boli of eastern Punjab and
- The Lavani of Maharashtra.

4. Puppetry :

In India four styles of puppetry have proved popular in different parts of the country :

- Sutradharika
- Rod puppets
- Shadow puppets
- Hand puppets

Form of indigenous communication :

Folk media : It includes festivals, plays and puppet shows, dance, song, story telling, poetry, debates. They are used primarily for entertainment, but they can also promote educational values and help in cultural continuity.

Indigenous organization and social gatherings : It includes religious groups, village meetings, irrigation associations, mothers clubs and loan association.

Deliberate instruction : Parents teach children, crafts people instruct apparentices, elders guide young people and adolescents undergo initiation rites. Many societies have traditional religious schools.

Record : Many societies keep records in written, carved, painted and memorized forms.

Unstructured channels :

It includes talk at home and at the well, in the field and on the road, in tea house and coffee house, in the chief's house and at the market and wherever people meet and talk. This communication is not organized or orchestrated but spontaneous and informal. Communication does not have to be intentional.

Advantages of using traditional media :

1. The traditional media are the closest to the hearts and minds of the masses, and so their appeal is at a personal, intimate level.
2. Their familiar format and content, as also the colloquial dialects used, make for clarity in communication.
3. Cross-cultural communication hurdles are not encountered here.
4. Rapport is immediate and direct, the barriers to communication almost non-existent.
5. The traditional media have their own method of communication. Their peculiar style may evolve from the local customs, traditions and various other social practices.
6. These pass the message in the common familiar language of the people.
7. As there is face to face interaction between the conveyer and the recipient, there is a scope for clarifications of doubts and acquisition of full information.
8. These media often appeal to the emotions of members of the audience, who identify themselves with the players.
9. The audience tends to have confidence in the messages because communication takes place in a form they are accustomed to, in their dialect and from people they often knows.
10. Audiences are seldom passive listeners or viewers, but tend to participate actively, thus influencing the performers. The Keertan and Alha (of U.P.) are examples of this.
11. The greatest advantage of the traditional media over the electronic media is their flexibility in accommodating new themes.
12. These media cater to small audiences at a time, but these audiences are so completely caught up in the folk forms that the effects on them are at a much deeper level.
13. The traditional art forms satisfy the innate desire for self-expression and also satisfy man's need for moral instruction combined with entertainment.
14. The traditional media have both entertainment and information value.

15. The traditional media are credible and culturally compatible.
16. The traditional media are inexpensive and informal.

Importance of different forms of indigenous communication :

Folk media :

Folk media are intimate with masses, rich in variety, readily available at low cost, relished by different age group and by both sexes, and have greater potential for persuasive communication, face to face communication and instant feedback (Rangnath, 1976). Folk songs can be both effective and efficient as a communication medium for a society which has a strong oral tradition. Folk songs can made social commentary to express feelings or thoughts, to arouse people to action or to insult those who misbehaved (Kidd et.al. 1976).

The Indian Government established a Song and Drama Division in 1954 for organizing entertainment programme with the help of local theatrical group on development themes. All India Radio also used folk media in its rural broadcasting in the form of a daily programme (Rangnath, 1976). Traditional media are important channel of communication as they employ the idiom of the people and the symbol which are readily intelligible to them, reach a part of the population that is impervious to the influence of mass media and demand active participation in the process of communication (Dissanayake, 1977).

Devdas (1978) stressed on the use of traditional media for effective education. Several traditional methods can be used in secondary school for teaching Home Science. Among these, story telling, folk songs and dances are outstanding to augment in imparting education in an interesting and effective manner. Patron (1981) has revealed that story telling is a very effective tool for education and socialization of children. It has been used in the "Pabasa" project of the Nutrition centre of the Philippines which revives story-telling to its traditional popularity and to use it to effect development information goals.

Kumar (1981) found in his pilot study on comparative power of puppetry and a documentary film in a two villages near Delhi that the cheaper traditional medium could be as effective as films. Women in particular responded more favourably to the puppet shows than to films. Malik (1985) stated that India has the tremendous varieties of folk singing styles. All these styles are capable to pass socially significant development messages. Folk songs have often been used in India to stir up emotions against the British during pre-independence period, to foster a sense of national pride and integration and to enthuse people about various projects.

Malhani (1995) stated that the Government of India has been the biggest user of traditional media. Doordarshan has been the pioneering institution to draw upon folk music. The Radio, Doordarshan dramas and theater have integrated various folk theatre forms. Jatra, Tamasha, Nachas, Bhavai, Nautanki, Yakshgan, Therukuthu, Kavi Sammelans and Mushairas are used by various stations of All India Radio. Intodia and Upadhyay (1991) found that traditional media have remarkable impact on the rural society because of their acceptable idioms, functional significance, emotional touch and entertainment component.

Saxsena (1993) reported that traditional media are more effective in rural areas because the techniques are simple and easy. Traditional media can more effectively stir the feeling and affect the attitude of illiterate masses. Murthy (1994) found that folk arts are laying a meaningful role in rural areas, in educating the rural people about the consequence of social evil like alcoholism, illiteracy, population-explosion, untouchability, superstition, communialism, malnutrition and insanitation, dowry, satipratha and so on. Samar (1997) reported that the traditional media like puppets always have powerful impact and are used not only for entertainment but also for educational purposes in school and the society.

Indigenous organization and social gatherings :

Traditional communication forum such as the market place, tea or coffee houses, pubs, transportation system and village meetings should be used to reach rural audience (Patron, 1976). Village wells, shops, temples etc. are important meeting places and useful to display educational and motivational material or use audio/audio-visual and other media to disseminate massages (UNICEF, 1986).

Tribal market maintains a network of socio-cultural ties among the tribal people and gives them a common base for regional ethnology. He further reported that "Banari inter-tribal" market is located in Ghotanagpur as an agent of cultural change and it serves to bring together a number of communities like the agriculturist Oraon, hunter and food gather Birhor, horticulturist Birjia and Kissans, the ironsmelter Asur, the weaver chik Baraik, the basket maker mahali, the non tribal Hindu Baniyas etc. living in the hinterland of Banari (Sinha, 1968). Traditional midwives of Malaysia, Phillippines, Thailand, Mexico and Indonesia have very high credibility about health and related matter (Rogers, 1975). Midwife is the best communicator in many parts of Asia (Chalkley, 1976). Weekly market plays an important role in the life of the tribal people. The tribal market brings together people from different ethnic group for not only economic but also secular and religious activities in the tribal region. They further found that the impact of weekly market on their traditional life has shown an attitude of accepting innovation. The market on their traditional life has shown an attitude of accepting innovation. The market is the most powerful channel of communication in the tribal region (Vidyarthi and Rai, 1976).

The Comilla project in Bangladesh was built into traditional communication apparatus such as village meeting. He further stated that village meeting setup in Indonesia has been used for development campaigns (Syed, 1976). Folk media give more widespread awareness of information than the public address system (Verghese and Bhanari, 1977). Africa has a number of unique traditional communication technique, among them are the "Taking Drums" and "Taking Horns" which can be used for achieving development communication goals. The early missionaries on continents learned to use the drums and horns to convey their religious messages (Ludovick, 1978) The "Fiestas" (village festivals) have served as an occasions for political education. Specific socio-economic issues have been scripted into folk debates, desirable value system have been woven into folk drama and sloganeering has been introduced to take advantage of the people's propensity for proverb and other forms of rhythmetic witticism (Patron, 1981).

The social functionaries like barber's wife, Dai, Washer-women, bangle sellers etc. are important source of information to village women. They may be utilized as health workers for information dissemination on health and family planning (Sharma, 1996). Non-political organizations are better way to reach people and members of these organizations should be trained and used for different type of development activities like health, hygiene, and other social works. Artisans, well, bazaar complex, haat, fair etc. are crucial point for initiating development communication (Keshava, 1997). Traditional institutions, traditional healers, traditional leaders, traditional taboos, sacred species and sites have active role in conserving the forest and various strategies must be combined including the use of traditional institution (Kajembe et.al. 2003).

Traditional leaders :

The most effective ways of convincing rural audiencesis to enlist prominent village personality to spread message (Mahoney, 1975). Traditional as well as religious leadership can be utilized for easy flow of information and for successful implementation of programme after taking them into confidence (Keshava, 1997). Traditional laders play an important role in the lives of rural people and have a great influence in decision making and involving everyone in the communities (Peires, 2000).

Conclusions :

Traditional media may compete for attention with modern media such as radio, television, cassettes or video, and in some instances have lost the battle. People do not like to be considered traditional in their use of media, but in some situations different media forms can co-exist with each other. Folk and modern media do not have to compete, but may complement each other. Folk and modern media can be shown on television, and folk songs with a development theme can be broadcast on radio. This often is done in Indonesia, partly because the government sees it as a way of promoting national unity. A Brazilian agricultural extension service commissioned a popular folk singer to write and perform folk songs which incorporate technical messages about fertilizer use (Van den Ban and Hawkins, 1998). Regarding comparative effectiveness of traditional media Namdeo and Mishra (2005) have reported that traditional channel of communication (Munadi) is more effective than interpersonal contact by KVK worker and newspaper in disseminating messages of local importance in the tribal village. The Union Bank of India and the Life Insurance Corporation have used the medium of puppetry with great effectiveness in U.P. (India) to arouse the interest of rural folk in bank savings and LIC policies (Kaul, 1999). The State Bank of India, Pantnagar branch has also used puppetry effectively in Pantnagar Farmer's Fair to arouse the interest of the visitors in bank savings and loans.

References :

1. Devdas, R. P. (1978) Method of Teaching Home Science. NCERT Publications. New Delhi (cited by Tiwari et. Al., 2003).
2. Dissanayake, W. (1977) New wine in old bottles : can folk media convey modern messages? Journal of communication, spring : 122-124.

3. Intodia, S. L. and Upadhyay, R. (1991) Puppetry : A traditional media for rural communication Maharashtra Journal of extension Education. 10(2) : 191-195 (cited by Chapke and Sonatakki, 2002).
4. Kajembe, G.S.; Luoga, E. J.; Kijaji, M.S. and Mwaiprpo (2003). The role of traditional institution in the conservation of forest resource in East Uasambara, Tanzani. International Journal of Sustainable Development and world Ecology. 10(2) : 101-107.
5. Kaul, P.N.(1999) Traditional Media for Development Support Communication. Lecture delivered during Summer School on Media Production Skills for Development Support Communication held at Pantnagar from 10-30 June.
6. Keshava (1997) Indigenous communication system in Buxa Tribes : A case study. Tui, 3(17) : 1-13.
7. Kidd, Rose and Martin Byram (1976) Folk Media and Development : A Botswana case study. Paper printed by the Botswana Extension College (cited by Wang and Dissanayke).
8. Kumar, B. and Hansara, B. S. (2000). Extension Education for Human Resource Development. Concept publishing company, New Delhi.
9. Kumar, K. J. (1981), Mass Communication in India. Jaico publishing house, New Delhi (cited by Tiwari et.al., 2003).
10. Ludovick, A. N. (1978) From Talking Drum to Talking Paper. Index on censorship. May-June : 29-32 (cited by Lent, 1982).
11. Mahoney, H. (1975) Notes on Rural Communication Program. Malaysian Government's Field Services Division Department of Information. July 15 August, 20. Unpublished paper. Temple University (cited by Lent, 1982).
12. Malhani, P.N. (1985) Traditional media for effective communication media, ministry of I and B. India, pp : 165-172 (cited by Chapke and Sonatakki, 2002).
13. Malik, M. (1985) Traditional forms of communication and the mass media in India. Communication and society. UNESCO. Paris : 25-26.
14. Mundy, P. and Compton, L. (1991) Indigenous Communication and Indigenous Knowledge. Development communication Report, 74(3) : 1-3.
15. Murthy, J. s. (1994) National Communication Policy and Folk Media for Development. Communicator, January-March : 32-33 (cited by Chapke and Sonatakki, 2002).
16. Namdeo, M. and Mishra, O.P. (2005) Indigenous and Exogenous Channels of Communication in a Tribal Village. Indian Research Journal of Extension Education, 5(2&3) : 61-63.

17. Patron, S. Josefina (1976) Communication and National Development. Silliman Journal, Second Quarter : 123 (cited by Lent, 1982).
18. Patron, S. Josefina (1981) Role of Tradition Media and Art Forms in Films. Media Asia, 8(3) : 191-195.
19. Peires J. B. (2000) Traditional Leaders in Purgatory Local Government in Tsolo Qumbu and Portst. Johns. African Studies. 59(1) : 97-114.
20. Rangnath, H. K. (1976) A Probe into the Traditional Media Telling the People Tell Themselves. Media Asia, 3(1) : 25.
21. Samar, D. L. (1997) Use of Traditional Media for Communication. Reading in traditional media vol. (1) IIMC, New Delhi.
22. Saxsena R. P. (1993) Rural communication problems and perspective. Communicator. Jan-March : 22-25.
23. Sharma, N. (1996) Channels of Health Communication in Rural India. Interaction. 14(2) : 108-119.
24. Sinha D. P. (1968) Culture change in Inter tribal market. Asia publishing house, Bombay.
25. Syed, A. Rahim (1976). Communication and rural development in Bangladesh. Honolulu : East-West Communication Institute (cited by Lent, 1986).
26. UNICEF (1986). Micro planning and management course in EPI : Users manual for multipurpose workers part II UNICEF, regional office for south central Asia, New Delhi (cited by Sharma, 1996).
27. Van den Ban, A. W. and Hawkins, H. S. (1998) Agricultural Extension. CBS Publishers and Distributors, New Delhi.
28. Verghese T. and Bhandari, S. V. (1977) Folk media as a channel of communication : A study of "munadi" in a rural area. Journal of family welfare 23(4) : 33-39.
29. Vidyarthi, L. P. and Rai, B. K. (1976) Economic system of the tribals. The Tribal culture of India. Concept publishing company, New Delhi.
30. Wang, G. and Dissanayake, W. (1982) The study of Indigenous Communication System in Development : phased out or phasing in. Media Asia, 7(1) : 3-9.

Chapter **37**

Participatory Research- a Tool of Community Communication

Pandey K. N. and Dubey V. K.

The rural areas are critical to the life of all people. Most of the Earths' natural resources are in rural areas. It would be a mistake to assume that the only purpose for rural development is for the poor and disadvantaged people who have been left behind in a rapid advance towards a technological world of assumption. These rural areas which are so critical to people are inhabited by most of the world's population. Because of lack of appropriate technology and in many case because of the application of inappropriate technology; life is neither comfortable nor it makes efficient use of local resources. Use of local forest wood for fuel, loss of grains through improper storage and suffering of people due to polluted drinking water are some of the examples.

Only fulfilling the basic needs by providing food, clothing and shelter is not the solution to the problem, but the mechanism by which basic needs are met may promote a dynamic development of the process.

At every step of the development the user's participation is of much more important than the product or process put to use.

Local participation is not the only new criteria by which rural development need to be judged. It is equally important that problem be approached holistically taking in to account the full range of human and community potentials.

Participatory Research (PR) :

Participatory research is a systematic process of involvement of people for the analysis of their own situation and finding out solution for improving their present condition." It can be grouped into two parts.

(A) **Explorative research** : is an effort to introduce innovation not known to social system.

(B) **Amiliorative Research (AR) :** It emphasizes in the improvement of what people are doing (at present). It is an effort to make the good better (Compatible).

Characteristics of participatory research :

Like the hard-pan in soil, our social system is full of traditions impermeable for modernization. To break this tradition and bring improvement in the condition with the help and participation of the people may be the primary function of PR. For example, in traditional sectarian society like ours, if all women get a chance of self-introduction in a group of village women club, it will help them to see the similarities in their situation. This act of introduction, a very small act in itself, have tremendous effect on their confidence and create familiarity among themselves. Assembly of the group and discussion on certain problems will lead to collective and systematic analysis of the situation. Participatory research may be characterized as follows :

1. An activity carried out by group of people to understand their own situation for their own benefit.
2. It is a collective effort to bring desirable results taking into consideration the various aspects of problem in a collective manner.
3. It is an analysis that lead to action.
4. Understanding of personal realities and larger social realities, it will facilitate awareness raising about the connections between micro-aspects of their life and the real macro-aspects of society at large.

It is the involvement of people in decision making, development, ownership and control of programmes, products and process.

Participatory Research emphasizes that the application of research and its transfer should begin and end with the participants and involve interdisciplinary teamwork in all phases of continuous research/ diffusion process.

The success of participatory research lies in if team members of different discipline, function as equal partners with joint responsibility for the final outcome of the research. It is a series of goal achieving acceptable solutions to participant's problem which are linked in circular form by a number of activities.

Cyclic process of participatory research :

Participatory research is a cyclic process starting from grass-root (practitioners) to international level scientists. In this process, all those are involved, who equally participate in making the programme, executing and getting benefit out of it. It is a group activity. It has been depicted in figure 1.

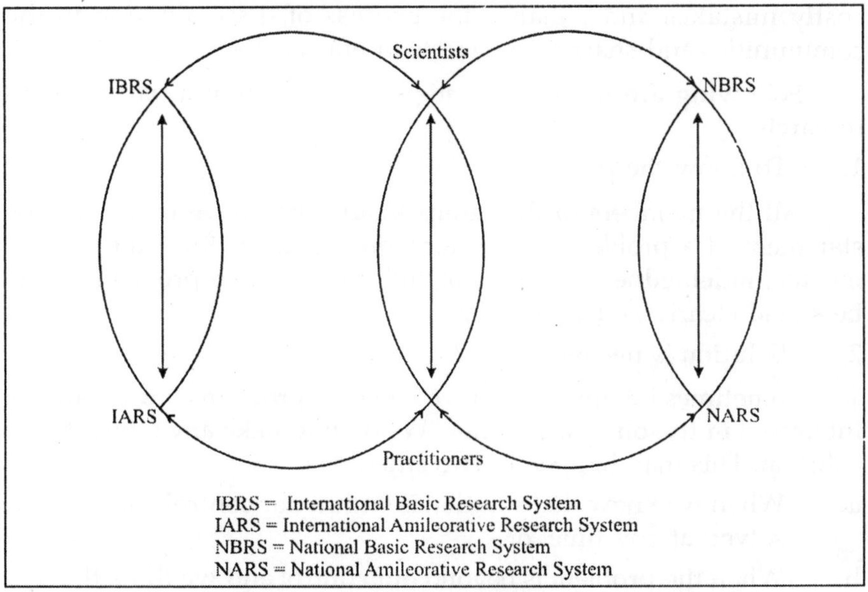

Fig. 1. : Cyclic Process of Participatory Research

The diagram shows as integrated system where each sector relates to the other and can influence them in turn. To be most effective each element of the system needs to be responsive to the others. It is timely to remind ourselves that any social system can not work successfully in isolation from other part of the system. If at certain level something is inadequate or obsure or delayed by maladministration or any other reason additional help may be procured from the other level for the welfare of the society.

This model comes together at the point of intersection and shows that the demand for participation will effect the support required and delivery system will be conditioned by the materials provided and so on. The limitation of the present model may be the least participation of practitioners in various systems.

There is need to consider the importance of participatory research at various levels of society. So far, it has remained confined to isolated efforts. The UN sponsored institutions, based on crop or livestock research do have made some attempts in this direction. However, the country based institutions are still shy. Therefore, we need to break the compartment in institutional and non-institutional efforts and involve the ultimate users in the process. It will save us from costly mistakes and enhance the process of dissemination in the communities and sharpen the facets of efforts also.

Following are the various steps in conducting a participatory research :

1. To know the problem :

All the members of the group should first agree on a common statement of a problem. Some common problems, for example, low income, indebtedness, sickness of children, etc. The problem should be stated clearly and concretely.

2. Solution is needed or not?

Sometimes it happens that we have the problems but we are not interested in its solution at a time. We do not make any efforts for its solution. This may happen in two situations.

a. When we know that solution is within our control and it can be solved at any time, or

b. When the problem is beyond our control and we think that it is wastage of resources and time because we can not solve it.

Therefore, the group should express interest in solving the problem.

3. To know the present status and additional information needed for the problem to be tackled :

All previous information (history) known to the group should be recorded about the problem. For example, if the problem is sickness of children, then they can each describe the sickness, its type, intensity and duration in the case of their own children. This helps in focusing on the problem in a concrete way.

After articulating and recording the thing known about the problem, one can begin to identify what is more they need to know about the problem. For example, how spread is the problem? Reasons for its occurrences in this form? Taking into consideration the same example of sickness of children, the group may need to know how many children in the area of operation are affected by the disease, why the sickness occurs, what are the way to prevent and cure the sickness etc.

4. The ways of collection of additional information :

This involves the steps of data collection. How to get the information, whose assistance is needed, who will do what and upto what time the information will be available?

5. Analyzing the available information :

Having obtained the information the group collectively analyses that information. It is important to note that analysis and reflection is undertaken as a collective process, not to be done by one person alone. This will help the group to understand, the possible causes and provide clues for solutions.

6. What actions are needed for possible solutions :

Various solutions to the problemare sorted out, their workability is assessed and the choice of solution is made. Then, for solving the problem a decision has to be made that what actions are to be taken, by whom, when, where and how. After this, actions, are taken.

7. Review of the actions :

Systematic assessment of the impact of the action taken should be made to see if the problem is fully or partially solved. The actions taken may sometimes create new problems. For example, if in a community there is need for safe drinking water and tap water is provided to solve this problem this solution to problem may create another problem of drainage in the lanes (which used to be otherwise dry). Now, a group exercise will be needed to work out agreed solution for creating a draining system in the community.

Generating acceptable technology (GAT Model) :

The fig. 2 emphasises the involvement of interdisciplinary teamwork in all phases of ongoing research and its transfer for the use of practitioners at the grassroot level. It is based on the assumption that the outcome of research should be applied and transfer should begin and end with the community. For official utilization of available resources a proper assessment of problem is essential. Through a process of interdisciplinary dialogue and interaction in the form of debate and constructive conflict the different facets of problems are brought together to arrive at a common definition. After this step, an attempt should be made by the team to develop potential solutions of the problem and proceeds for testing and adaptation. In cooperation with the participants potential solutions are evolved and compared with existing practices. The participants made a final decision after evaluation of technology under their own situation, using their own resources and management. The scientists monitor the technology time to time to determine how it is beneficial for participants and consumers.

For the best use of available resources, proper diagnosis of the problem is most important step. It may require the help of research scientist to make in depend observation. At this stage some constructive suggestion may come but at the same time statements creating confusion may emerge which might jeopardize the depth and importance of the problem for the time being. The next step becomes very important to sort out the common problems of the participants for which solutions are to be evolved. In the process of defining the common problem, interdisciplinary dialogue and interaction, generally initiated by debate, taking into consideration the socio-technical perspectives together, constructive suggestions are noted.

In this stage, community may put forth some solutions by itself that should also be thoroughly examined and if only improvement in the same is needed the possibility should be worked out. The research team should make all efforts to evolve some potential solution of the problems. It should be tested in the community situation for adoption. When the solutions are worked out, it should be finally evaluated by the scientist in cooperation with participants under the existing resources and management. Even at this stage the community may accept or reject the technology. If rejected, it gives further path for research to find out the reasons of rejections and may find ways to improve the performance and acceptability of the technology. If accepted the scientist monitor time to time to determine how it is beneficial to other communities.

Use of media in participatory research :

The rational underlying the use of communication to enhance popular participation is that people can only be motivated towards decision making if, firstly, they feel that they have been consulted by the planners during the design phase of a development project in other words, their felt needs and aspirations have been taken into account; and secondly, if they have information on which to base

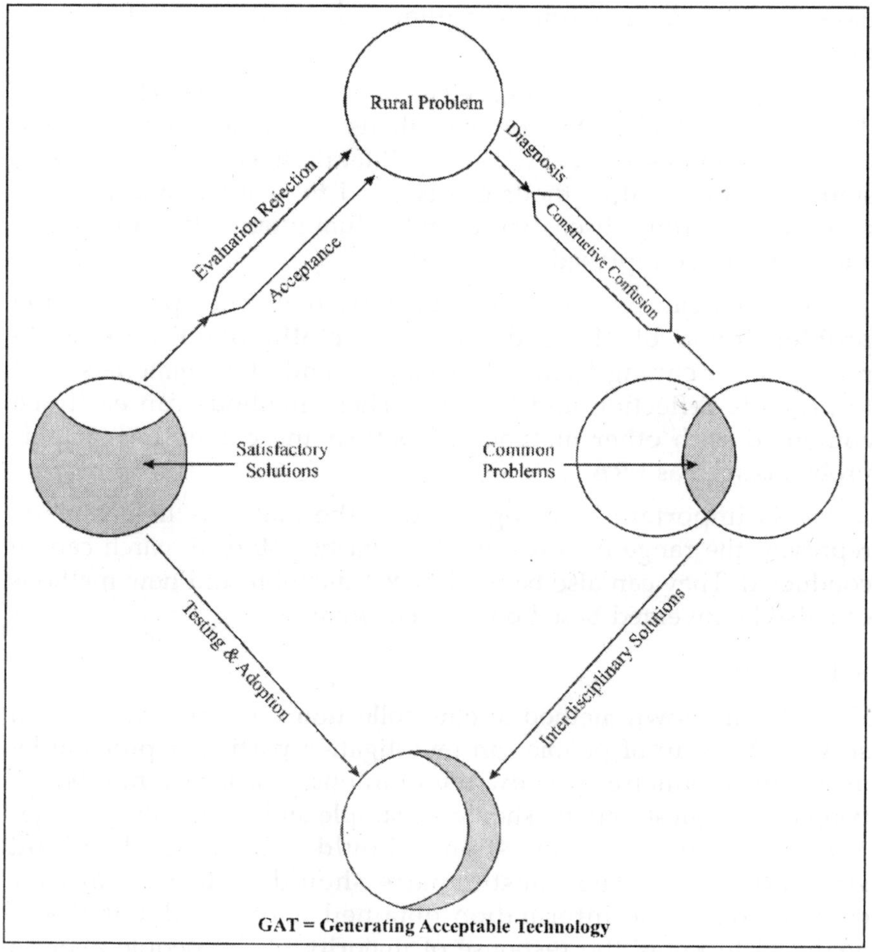

Fig. 2. : GAT Model

their decision. They also need a continuing possibility to making their opinions known as the project proceeds.

Communication techniques using such devices as CD/DVD/ Computer or audio tape recorders, photography, rural newspapers and the like can give life to a consultation process through a two way flow of information. They can also provide, in a comprehensible, audience tailored form, the information needed for decision making. Even TV and radio, normally one way channels, by imaginative use of audio and video recordings made in the field, among the target audience and by subsequent broadcasting to them and to peer groups can be used effectively. In addition, it is possible to achieve a feedback of audience reaction to authorities but this can also cause problems.

Sometimes the group engaged in Participatory Research may find itself uncomfortable with written methods. This is specially true with poor communities, most of whom are illiterate. In such circumstances, simple audio-visual methods can be used to facilitate analysis and reflection. A group discussion or individual interview can be taped and then can be replayed for analysis.

Women can be encouraged to draw their perceptions of the problem on a blackboard, paper or cloth. Black and white photographs, cutting from old magazines and other materials used to promote reflection and analysis. These methods can easily be combined with other methods to express the perception of their problems such as PRA.

It is important to recognize that the methods listed below represent the range of ways in which participatory research can be conducted. They can also be used in combination, and new methods can also be invented based on a given situation.

Self survey :

A well-known method of data collection is to carry out a small survey. A group of people can investigate a particular problem by using questionnaire to know the opinions of a larger number of people. The questionnaire should be simple and have only a few (8-10) questions. Each questions should be short, clear and straightforward. The questionnaire should seem like an easy conversation. The information obtained may be tabulated and analysed to know the response of majority of the people about a particular problem.

Group discussion :

A small number of people (10-25) come together to share experiences and decide for any action through group discussion on a particular problem. It is one of the most popular and easy method of combining data-collection, analysis and action planning. In a group discussion, contradictory opinions and points of view of various members can be openly discussed and analyzed. This facilitates a deeper probing and analysis of the problems as well.

The important consideration of group discussion is to insure the comfortable atmosphere in the group so that all may participate actively and listen to each other. If an individual perceives that he is not listened to, he may feel that his problem is no important and therefore, he is ignored, if the group is large, it can easily be sub-divided into smaller groups to facilitate free and easy discussion. This method can easily be combined with other methods.

People's camp :

A camp or shivir of workshop or training programme of people can be used as an important event to carry out participatory research. The period of the camp may be put over a day or two with common lodging and boarding arrangements. Generally, people from different villages, over a long distance come together.

In addition to education of people, it can provide an informal and free atmosphere. These camps provide an opportunity as well as reflection by the participants away from their daily pressure and facilitate development of feeling of solidarity among themselves. In a camp approximately 50-100 people can participate easily depending upon the facilities available at a point of time and place.

Research team :

For many problems chosen by the group, they may not have the amicable solution and it may require the help of scientists for technical advice. In these situations, members of the group join hands with outside experts in studying the problems and finding appropriate solutions for the benefit of the community. The scientists with their team extend their services and expertise to cater the needs of the group.

Visit to a demonstration (fact finding visit) :

It is always beneficial to visit an area where a similar problem has been solved for which the group is interested. The visit of another village can become an investigative tour to find out how other similar people have tried to solve the similar problem. It helps to feel confident as well. After the visit, this can be shared with other members of the community to solve their problems.

Folk theatre :

Indian folk theatre has an important role to play in the process of social change in India. The folk theatre media, low cost and indigenous and may prove better for focusing and solving rural problems.

Rural India comprises more illiterates than urban population. But they have tremendous potentialities in terms of picking up their own subject from their own life, viewing the issues and problems from their own working perspective and presenting the content in an appropriate form, elements of which are often from their folk background, Rao (1986) indicated that rural poor can communicate better and convince people by making their own theatre. He was of the view that they are highly potential and quite capable of making their own arts and particularly theatre, if they come out of their middle-class imposed conditions. He further opined that middle-class should resist the temptation of doing theatre for the poor and should help the poor to make their own theatre.

Indian folk theatre is full of exploitative stories of kings and priests tuned into Gods, picked up from different Hindu epics like Ramayana and Mahabharat. They have the "content" and the "form" agreeable to all. These people do have all the capabilities to bring all essential components of theatre meaningfully together to create and perform a play to the satisfaction of rural audience.

Analysis and interpretation of the problems can be attempted through drama and this media may be sued greatly for wider realization of the issues, its depth and content. This could be further used for transferring the solutions of selected problems in the society.

In some places, simply produced dramas have been used as an effective method of participatory research. The group member may identify a problem and create a skit or play on it in the presence of others. It is then discussed among audience to stimulate further analysis and action.

Puppets :

Puppets refer to any inanimate thing animated on stage to communicate any subject. All the six instincts (Communicative, dramatic, artistic, musical, inquisitive and constructive) basic to human learning are stimulated and get satisfied through puppet show. Kanitkar (1993) was of opinion that puppetry is a powerful audio-visual media for group communication. It is also a folk and people's art forum. Workers in voluntary groups who have an aptitude for line arts may easily pick up puppet-making. The average person of a community who by virtue of his work culture already possesses an above average handicraft ability, will also find it easy to pick up puppet-making. Puppet can be made out of easily available and cheap materials in the community like paper mach, cloth, cardboard, mud, saw dust, hay etc.

Puppets can make an impact, it properly used with the active participation of people information regarding community problem can be gathered and solutions can be communicated to people by proper interpretation. In addition to all above points, puppetry is an "INDIAN" medium of communication which attracts people of all age groups. In this way, the puppets can be sued by the local communities for highlighting their problems and related perceptions. It can be used to communicate the possible solutions also.

Documents :

Many books, reports, papers, registers, diaries and documents may contain useful information about a particular problem. Analysis of these documents is another method which may be of some use to solve the problems of people.

There are certain limitations of this method. For example, it is sometimes difficult to locate and obtain documents, but they do have factual information. Another limitation with documents is that they require literacy skills to analyse them.

Use of video in participatory research :

Proper development can easily take place in a local context with decentralized set-up in which people have scope for meaningful participation in controlling their own lives and environment. Unless the content of media originates at the community level and relates to their views, it is useless. In the present day world, people are interested to democratize media use at the small community level, so that people can present their problems and explore possible solutions in the context of realities of their situations. In this way, communities can derive the strength and confidence to solve their own problems.

Video is the most suitable medium for generating interaction. It has the ability to present the community in both moving image and sound for immediate reproduction. Unlike movies, it does not require processing and can be shown to the public just after filming. Video tapes can be used for promoting audio-visual literacy for motivation, attitudinal change, behaviour reinforcement, community participation and entertainment. Similar results even with more clarity and immediacy can be now used through computer system.

Audio cassettes :

Audio cassettes may be of immense use for participatory research. The problems of people can be recorded and put up before the scientist and heard. They may find out some solution of the problem and can be sent back to participants in community. It may work as two-way channel.

It is a cheaper device as compared to video cassettes and computers are quite effective for convincing the people. It may be further used for sharing the experience and problems of one community with other. Thus, it can strengthen the horizontal communication process.

Authors have tried to present the PR process at conceptual level with the help of two structural diagrammes. Some of the methods have been also described. They can also be used in combination and new methods can also be invented. In no way it rules out the possibility of PRA use. Rather it is desirable.

References :

1. Anonymous. (1986). With our own hand : Research for third world development : Canada's contribution through the institutional development research centre (1970-1985). Ottawa, Ontario, IDRC 246 C.
2. Arbab, F. (1984). Rural University : Learning about education and development book, published from international development research centre, 60 queens street, Ottawa, Canada.
3. Deniels, D. (1987). Evaluation in national agricultural research, Proceedings of workshop held in Singapore, 7-9 July, 1986. IDRC 254 C Ottawa, Canada.
4. Horton, D.E. (1984). Social scientists in agricultural research. Book published from international development research centre 60 Queens Street, Ottawa, Canada.
5. Kilvin, Joseph, (1971). Innovation in rural India. Cited in the paper of B.S.S. Rao. JI, Interaction, vol. 1, No. 8, pp. 231-37.
6. Kanitkar, H. (1983). Puppets for consientization – a few cases. JI Interaction, vol. 1, No. 3, pp. 204-215.
7. Rao, P.V. Putan Chandra, (1986). In search of an alternative theatres of illiterate rural masses. Cited from the book 'SAMPRESHAN' NCDC Publication, pp. 125-148.
8. Roy, Pradipto and Rogers, E.M. (1969). 'The impact of communication on development', Cieted from the book 'SAMPRESHAN', Published by NCDC, Varanasi, pp. 231-237.
9. Schramm, Wilbur. (1964). Mass mèdia and national development, UNESCO, Paris.

Subject Index